AF540143

BIOLOGICAL TECHNIQUES

BIOLOGICAL TECHNIQUES

By

Dr. P.R. Yadav
Lecturer
Department of Zoology
D.A.V. College
Muzaffarnagar (U.P.)

&

Dr. Rajiv Tyagi
Department of Zoology
M.M. College
Modi Nagar (U.P.)

DISCOVERY PUBLISHING HOUSE
NEW DELHI-110002

First Published – 2006

Reprinted – 2026

ISBN: 978-81-8356-105-1

© Authors

Biological Techniques

Published by:

DISCOVERY PUBLISHING HOUSE

4383/4B, Ansari Road, Darya Ganj

New Delhi-110 002 (India)

Phone: +91-11-23279245; 23253475; 43596065

Mobile: +91 9811179893 / +91 9871656464

E-mail: discoverybooksindia@gmail.com

orderdphbooks@gmail.com

namitwasan9@gmail.com

web: www.discoverypublishinggroup.com

Printed at:

Infinity Imaging Systems

Delhi

Preface

The present title "Biological Techniques" has been written for undergraduate, postgraduate students opting biological, pathological, biochemistry, biotechnology courses. It deals with the various types of fixatives, stains and techniques used in the fields just stated. It is basically a simple cookery book, but like all good cookery books this volume should remain on the bench until its useful life is over and should never become embedded in the library shelf. This book aims at an explanation of the process involved in the preparation of tissues for microscopical study. The object throughout has been to get away from the empirical and rule-of thumb methods that have so long held away in the field of microtechniques, and to replace them, so far as possible, by methods, based on actual understanding of the process involved, in terms of chemistry and physics.

It has been the constant endeavour of the authors to furnish maximum substance, keeping in view the limitations of size of the volume. Efforts have been made to condense the matter as far as practicable. The book features both a text and a laboratory guide. It is hoped that this book will not only meet the requirement of Indian students but will also be useful as a guideline to the teachers in their teaching.

There can be no claim to originality except in the manner of treatment and much of the information has been obtained from the books and scientific journals available in different libraries.

The authors express their thanks to their friends and colleagues whose constant inspiration have initiated them to bring out this book.

The authors express their gratitute to Mr. Wasan and staff of M/s Discovery Publishing House for their whole hearted co-operation in the publication of this book.

Authors

Contents

1

INTRODUCTION

There is only one really reliable criterion by which we can determine whether the image that we see with the microscope is a good representation of what existed in life, and that criterion is comparison with the living cell. We have nowadays several excellent ways of overcoming the natural transparency of this object. Phase-contrast and interference microscopy are the chief methods, and vital dyeing, though at the moment partially eclipsed, remains potentially as illuminating as ever. Since we possess these reliable means of obtaining knowledge, it may reasonably be asked why we need also the elaborate procedures that form the subject-matter of this book.

Many kinds of cells cannot be isolated for separate study while still alive. Such cells can only be examined in permanent preparations. It is a great advantage to be able to cut tissues and cells into thin slices. The relations of the cells to one another and to the intercellular matter are much better shown in this way than when cells are teased apart for direct study while still alive. The living cell of many-celled organisms cannot be sectioned effectively. Even if a slice could be cut, it would not remain alive nor retain its form. Tissues and cells need to be stabilized in structure and held firmly while being sliced. In fact, they require to be 'fixed' more or less in their living form and then embedded in some solid material before they can be sectioned. When the process of fixation has been carried out, nearly all their parts can be dyed in contrasting colours, and this makes observation much easier. Vital dyeing, invaluable though it is, is applicable only to particular components of the cell. Again, few histochemical tests are applicable to unfixed tissues. Beyond all this, it is a convenience to be able to store away permanent preparations, ready for instant examination at any time.

For these reasons it is usual to treat a piece of tissue with a fluid called a fixative; to embed it after fixation in some solid medium, such as paraffin wax, that will hold its constituent parts in the right relation to one another during sectioning; then to section it; next to dye the sections, often in two or more contrasting colours; and finally to mount the dyed sections in a medium that renders them transparent.

In this book we are concerned with the principles underlying fixation, embedding, dyeing, and mounting. We are not immediately concerned with the application of histochemical tests, a subject to which another volume in this series of Biological Monographs is devoted. Nevertheless, the present work is largely chemical in outlook, and the preparation of tissues for study follows the same general routine whether a histochemical test is to be applied or not. It is therefore hoped that the book may be useful to those whose main interests are in histochemical or cytochemical analysis. We shall study *what happens* when we fix, embed, dye, and mount tissues (and especially the cells that they contain).

If a piece of tissue is cut out of a living or recently dead organism and no special care is taken to keep it alive or maintain its structure, it will soon undergo marked changes. If left in the air, it will lose water by evaporation and shrink; if left in a fluid, it is likely to undergo osmotic swelling or shrinkage. If these distortions are prevented, it will still be subject to attack by bacteria and moulds. Even if these are excluded by asepsis, the tissue will gradually fall to pieces by self-digestion or 'autolysis'. Cells contain enzymes, collectively known as 'cathepsin', capable of dissolving their own protein constituents when they die. These enzymes (two proteinases, a carboxypeptidase, and an aminopeptidase) show a remarkable resemblance to those of the digestive tract.

To preserve a piece of tissue one requires a fluid that will not shrink or swell or dissolve or distort; will kill bacteria and moulds;

and will render the autolytic enzymes inactive. Such a fluid is *a preservative. A fixative* must do everything that a preservative does, but in addition it must modify the tissues in such a way that they become capable of *resisting* subsequent treatments of various kinds. Of these treatments the ones that are most likely to cause damage are embedding, sectioning, and mounting (though the latter is not damaging if the tissue has already been severely shrunken while being embedded). Apart from the partial protection it gives against damage from these processes, fixation usually makes many tissue-constituents (especially chromatin) readily colourable by suitable dyes.

Not every tissue-constituent requires fixation, and some are not capable of being fixed. Chitin, cellulose, starch-grains, scleroproteins, amorphous silica, and certain inorganic crystals are examples of hard, stable substances, scarcely or not at all subject to distortion or decay. Other substances, such as the soluble sugars, cannot be retained in their natural sites in the tissues by any fixative. Glycogen is the only soluble carbohydrate that is at all frequently retained in finished microscopical preparations.

Though usually soft or indeed liquid, triglycerides do not require fixation if lipid-solvents are avoided in subsequent treatment. Some fixatives have no effect on lipids, but this does not by any means make them useless, for neither the cell nor intercellular material is held together wholly by lipids. If, however, the common proteins are not stabilized, tissues and cells fall to pieces. The essential function of fixation is the stabilization of the protein part of the framework of the cell.

Fixation can be achieved by heat, but this method is seldom used except for blood-films. It tends to cause distortion, but presents the advantage that nothing is dissolved out of the cell.

Most fixatives are solids used in aqueous solution, but some are organic liquids that can be used without dilution. The number of chemical compounds that are useful as fixatives is very small. The seven unmixed or 'primary' fixatives chosen for mention in this book are listed below. All fixatives fall easily into two groups, according to their obvious reactions with soluble proteins. Some of them, when mixed with a solution of albumin, produce a coagulum: others do not. The ones that produce a coagulum in a test-tube transform protoplasm into a microscopical spongework.

Against each fixative in the list below is marked its 'standard concentration'. This is the concentration at or near which it is commonly used in fixation. Throughout the book, except where the contrary is distinctly stated, it is to be understood that when reference is made to one of the seven selected primary fixatives, it was used at the concentration shown in the list (or at a concentration so close to this that no difference in result could be anticipated).

COAGULANTS

Ethanol **(ethyl alcohol). C_2H_5OH.** A light, colourless fluid, miscible with water in all proportions. Standard concentration, undiluted (absolute).

Mercuric chloride. **$HgCl_2$.** Colourless crystals, soluble in water at about 7%. Standard concentration, saturated aqueous solution.

Chromium trioxide. **CrO_3.** Brownish-red crystals, giving a strongly acid solution in water, in which it is extremely soluble. Standard concentration, 0.5% aqueous.

NON-COAGULANTS

Formaldehyde. **HCHO.** Colourless gas, very soluble in water. Standard concentration, 4 % aqueous.

Osmium tetroxide. **OsO_4.** Pale yellow crystals, soluble in water at about 7 %. Standard concentration, 1 % aqueous.

Potassium dichromate. **$K_2Cr_2O_7$.** Orange-red crystals, giving a weakly acid solution in water, in which it is soluble at about 10%. Standard concentration, 1.5% aqueous.

Acetic acid. **$H_3C.COOH$.** Colourless liquid, miscible with water in all proportions. Standard concentration, 5%.

Since each of the primary fixatives has its virtues and defects, it is usual in practice to mix two or more of them together. Their effects on tissues can only be understood when those of their components are known. Further, the proportions of the primary fixatives in the mixtures have been very arbitrarily chosen. For these reasons, any scientific account of fixation must start with the primary fixatives and be concerned mainly with these.

Small pieces of tissue are best fixed by direct immersion, since this brings the fixative most rapidly into contact with the cells throughout the piece. When it is necessary to fix a piece of tissue a centimetre or more thick, it is best to inject the fixative through a blood-vessel in order to send it quickly to all depths. This method of 'perfusion' has the disadvantage that the total amount of fixative that can be contained in the blood-vessels is usually small, and nothing outside the vessels is fixed until the fixative substance has passed through their walls into the intercellular spaces and thus reached the cells.

The purpose of fixation is usually to stabilize the tissues so that they retain as nearly as possible the form they had in life, but clearly it is not the purpose to leave their chemical composition unchanged. This would indeed be the negation of fixation, for the cells would be still alive or like recently dead ones. The purpose is to change the chemical composition in such a way as to confer structural stability. This fact by no means makes histochemical studies impossible, for certain tissue-constituents may be left unaltered, and others only altered in part.

The structural formula for amino-acids in general is conveniently written in the way because such formulae may easily be joined together to make a protein chain. The letter R in the formula represents any of the radicles that distinguish the various amino-acids from one another. In forming a protein, the amino-acids condense together with elimination of one molecule of water at each peptide link $\left(\overset{\overset{\displaystyle O}{\|}}{-C}-\overset{\displaystyle H}{N}-\right)$.

In the formula representing part of a protein chain, certain conventions have been adopted. The radicles characteristic of each amino-acid are written on the right of the formula in each

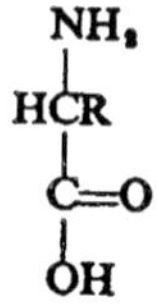

NH_2
HCR
C=O
OH

General formula for amino-acids

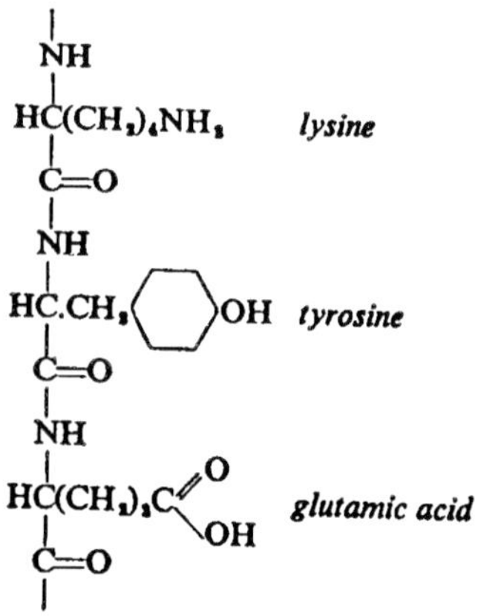

NH
$HC(CH_2)_4NH_2$ ***lysine***
C=O
NH
$HC.CH_2C_6H_4OH$ ***tyrosine***
C=O
NH
$HC(CH_2)_2COOH$ ***glutamic acid***
C=O

Three amino-acids as part of a protein chain

case, although in fact they project in different directions; and the repeated part or backbone of the formula is shown as straight, though in nature it is folded, either to form a simple zigzag or in more complex ways. These conventions will be found to simplify the explanation of the reactions of proteins with fixatives and dyes.

The first requirement of a fixative is that it should not be proteolytic. Any substance that breaks the peptide links and sets free soluble amino-acids is the opposite of a fixative. The changes produced by a fixative must tend in the other direction, towards stability.

One may broadly distinguish two kinds of fixation of protein, additive and non-additive. In the former, certain atoms of the fixative combine chemically with some part of the protein and remain in combination. In the latter, no such obvious addition of atoms occurs. Ethanol is a non-additive fixative in the sense that none of its constituent atoms joins itself to protein, so far as is known. The 'nature' of the protein is, however, profoundly changed by its action, and the substance is therefore said to be 'denatured'. The most obvious change is loss of solubility with resultant coagulation.

When a denaturing fixative is added to a solution of albumin, coagulation usually follows so quickly that it appears to be instantaneous. In fact, however, denaturation can be shown to proceed by stages. The first effect is an increase in reactivity; there then follows a flocculation or aggregation into minute particles, which are soluble in weak acids and alkalis; finally the flocculi join into a clot, soluble only by reagents that cause proteolysis. Flocculation and clot-formation are both included in the term coagulation. Throughout all these changes the backbone of the protein remains unaltered. Some authors mean by denaturation almost any change in a protein that increases reactivity but does not destroy the backbone. In this book the term will be used in a more limited sense to indicate a marked change involving the relation of the protein with the surrounding water. This change results in coagulation.

The chief hydrophil groups of proteins are the $-NH_2$ and

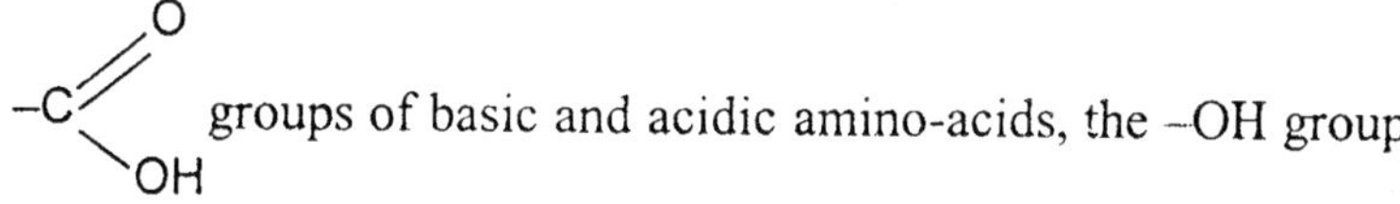

groups of basic and acidic amino-acids, the –OH group of tyrosine, and the $>C=O$ groups of the backbone. Whether a particular protein be dispersed as a sol or held together by cross- links to form a gel, it will ordinarily be related to water molecules by such groups as these. When a protein is denatured, the relations with water are somehow disturbed. It may be supposed that ethanol, a dehydrating agent, competes with the protein for the water, and that the active groups, now no longer associated with water, are more free to make fresh bonds with one another. The formation of fresh bonds, so close as to exclude much of the water that formerly lay between protein chains, would cause coagulation. Interfaces between 'dry' protein and the surrounding fluid would thus be formed; and since these would scatter light, the transparent sol or gel would change into a white coagulum.

Various reactive groups that were present in the unaltered protein in a latent condition are unmasked by denaturing and now respond to tests for their presence. This applies to various ionizing side-chains of the constituent amino-acids, especially the -SH of cysteine, the -S-S- of cystine, the phenyl of tyrosine, and the indolyl of tryptophane. The cause of this increase in reactivity is not known, but it may be a straightening of much-folded backbones and consequent exposure of previously-hidden side-groups. The protein is now more accessible both to dyes and to digestive enzymes. The specificity of the original protein is at the same time generally lost. This is most evident if the protein is or forms part of an enzyme.

Proteins can be denatured by very diverse means. Heat alone is effective, if water be present; in the complete absence of moisture, however, even a temperature of 100° C does not suffice to denature. Exposure to ultraviolet light or ultrasonic waves, subjection to very high pressure, and extension in extremely thin films can cause the denaturing of proteins, but these methods are not ordinarily used in microtechnique.

In additive fixation a part or the whole of the fixative molecule adds itself to the substance that it fixes, by making ionic or covalent links. In the fixation of proteins, the link is usually with the side-groups of one or more particular amino-acids. Additive fixation is sometimes coagulant, sometimes not.

An additive fixative need not necessarily alter the reactions of a protein very profoundly. For instance, it might act only on tyrosine side-groups, and if these were sparsely represented in a particular protein, most of the amino-acids might be able to retain their character and show their usual responses to reagents. The rest of the protein might undergo changes similar to those that occur in non-additive fixation. Thus there need not be a very sharp distinction between the two kinds of fixation, so far as the protein as a whole is concerned. Nevertheless, it is convenient to restrict the term 'denaturing' to cases in which the protein does not undergo chemical combination.

Different primary fixatives penetrate into tissues at different rates. Other things being equal, it is desirable that a fixative should penetrate quickly, so that the tissue may be stabilized in structure before autolysis has damaged it. In designing fixative mixtures it is helpful to know the relative rates of penetration of the components, for ideally one component should not outstrip another.

To study the rate of penetration of fixatives, it is best to begin by using a homogeneous protein gel in place of a piece of tissue.[120,21] A suitable gel may be made by dissolving gelatine at 15 % in warm white-of-egg. This material may be used as a crude model of cytoplasm, for its refractive index (and hence its protein-content) is about the same. While still warm it can be drawn into glass tubes, and the gel then allowed to set. If one end of the tube is inserted into a solution of a coagulant fixative, the rate of penetration can easily be noted, since the material becomes white and opaque on coagulation. To find how far a non-coagulant fixative has penetrated, it is only necessary to turn the tube upside down in a bowl of warm water, for the unfixed gel then runs out, while the fixed material remains in the tube. It will be understood that what is measured in these experiments is the distance through which the substance under test has moved at the concentration that is just sufficient to fix the gel.

Fixatives penetrate into the gel rapidly at first and progressively more slowly as time goes on. It has been shown that nearly all fixatives move forward in accordance with the laws of diffusion. If d is the distance penetrated in time t, and K is a constant depending on the fixative used, then

$$d = K\sqrt{t}.$$

Distance may be measured in mm and time in hours. K then represents the distance in mm through which the fixative has moved forward in one hour, at a concentration sufficient to fix the protein.

It is convenient to transform the equation just given into its logarithmic form,

$$\log d = \log k + \frac{1}{2}\log t.$$

This, when graphed, will necessarily give a straight line. If a set of figures for d and t, obtained in an experiment, are converted into logarithms and marked on a graph, it will at once be seen whether they lie approximately on a straight line. This is so with nearly all fixatives. Where the straight line crosses the ordinate line, log t has become zero and log d = log K. One can therefore read off the value of log K on the ordinate line, and the antilogarithm of the figure obtained is therefore K. In the case of mercuric chloride penetrating into gelatine-albumin gel, the K-value is *2.27*. Formaldehyde shows the highest K-value (3·6) among the seven selected primary fixatives. Chromium trioxide is slow *(K = 1·12)*.

It is convenient to make the units of the abscissa-scale (log t) one-half of those of the ordinate-scale (log *d)*. Then, if the fraction before log t in the logarithmic equation is ½ (that is, if *d is* proportional to the square root of t), the slope of the line will be 45°. This is so with most fixatives.

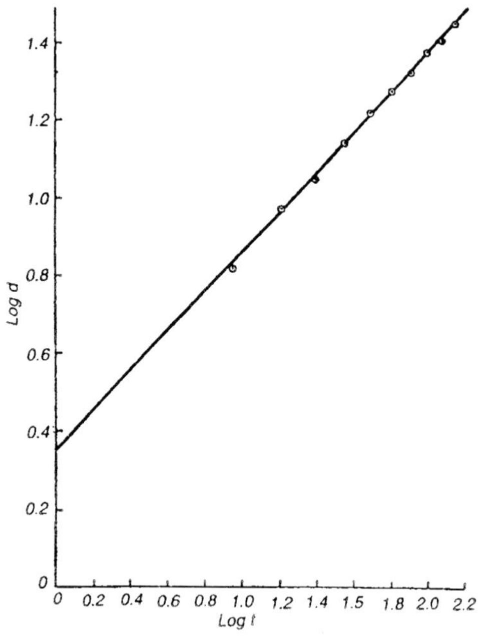

Figure 1.1 : ***Graph showing the rate of penetration of mercuric chloride (at standard concentration) into gelatine-albumin gel. t is time in hours; d is the distance penetrated in mm.***

Few tissues are sufficiently homogeneous to give clear-cut results in tests of rate of penetration. Liver, however, is suitable, and certain coagulant fixatives leave a clear indication of how far they have moved into this organ, since the fixed tissue is whitish and opaque. Experiments with mercuric chloride show that here also the laws of diffusion are obeyed. The K-value of mercuric chloride diffusing into liver is 0·84. So far as is known, all fixatives penetrate much more slowly into liver than into gelatine-albumin gel, perhaps because they are held back by the lipids in the various membranes of the cell. There is, however, a general agreement in the results obtained with gels and liver. Thus chromium trioxide penetrates much more slowly than mercuric chloride into both.

It is important to realize how rapidly the rate of penetration of fixatives falls off with time. With a K-value of 0·84, mercuric chloride penetrates into liver a distance of 20 μ (the diameter of a large cell) in just over two seconds. That is to say, it penetrates this small distance at the rate of 36 mm an hour; but actually it only penetrates 0·84 mm in one hour, and it would take 77 days to penetrate 36 mm. These facts emphasize the importance of using small pieces of tissue for fixation. Pieces several cm thick are only adequately fixed in their external parts, unless they are perfused with the fixative.

Instructions are often given as to the time during which a particular fixative should act, but it is often safe to disregard these. For optical microscopy it is convenient as a general rule to fix for 6 hours or overnight, though short fixation (about 3 hours) is best with Zenker's fluid, and even less often *suffices* for the pieces used in cytology, which are seldom more than 3 mm thick. In electron-microscopy osmium tetroxide is the most usual fixative. There is some evidence that proteins may eventually be dissolved by this fixative, after having been well fixed at first. Pieces about 1 mm thick are used, and it is quite usual to fix only for an hour or two, or even for less than an hour.

The structure exhibited in a fixed microscopical preparation is almost always to some extent artificial, for no fixative is perfect. Artifacts are of two kinds, extrinsic and intrinsic.

Extrinsic artifacts are those formed of material brought in by the fixative and deposited in the tissues. One may take as an example the black granules often seen in material fixed in mercuric chloride solution. Artifacts of this kind can usually be avoided. They are commonly caused by the reaction of the fixative left in the tissues with fluids in which the tissues are subsequently immersed. An extrinsic artifact, once formed, may often be removed by the use of special solvents.

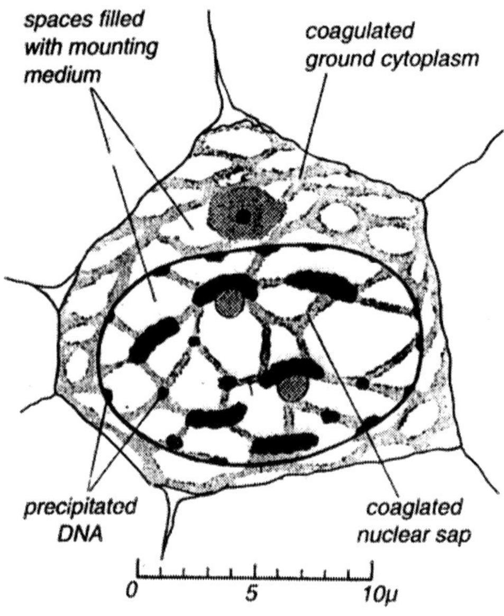

Figure 1.2 : ***A typical spermatogonium, as seen in a routine microscopical preparation (diagrammatic). The cell has been fixed in a coagulant fixative, embedded in paraffin, dyed, and mounted.***

The intrinsic artifacts are the distorted structures of the tissue-constituents themselves. Very great changes are undergone by a cell when it is fixed by a coagulant fixative, embedded in paraffin, sectioned, dyed, and mounted; and fixation itself is the chief cause of these changes. What happens may be studied by comparing fig. 4 with figure 1. The cell is irregularly shrunken, and the ground cytoplasm is transformed into a coagulum consisting of a spongework of protein strands. No part of the original substance of the cell is left in the meshes of the network, which are filled only with the mounting medium. The size of the meshes depends on the particular fixative used. They are often even larger than those shown in the diagram. The coarseness of the coagulum is not always realized, because the strands cross one another at different depths in the thickness of an ordinary section, so as not to leave any very obvious gaps, and rather feeble dyes are generally used for this part of the cell; but if a crudely coagulant fixative is used, and very thin sections are strongly dyed with iron haematein, the gaps become obvious.

No traces of the mitochondria or lipochondria remain. The centriole, if present, is intact; the idiozome sometimes but not always persists. The nuclear sap has been coagulated, like the ground cytoplasm. The heterochromatic segments of the chromosomes are still visible and strongly dyed, but the rest of the DNA has been thrown down in the form of an irregular precipitate scattered here and there on the strands of the coagulum and on the inside of the nuclear membrane. The nucleoli retain their form but are somewhat shrunken.

The non-coagulant fixatives, especially osmium tetroxide and formaldehyde, cause much less initial change in structure than the coagulants. Indeed, while the cells still lie in these fixatives they often look remarkably lifelike. Embedding in paraffin, however, usually causes serious distortion.

Ideally a fixative would leave the volume of a piece of tissue and of its constituent cells unaltered, and resistant to alteration by fluids in which it was subsequently placed. Many fixatives shrink tissues; some leave them scarcely changed in volume, while others swell them, but in any case the tissue is liable to subsequent shrinkage.

The results of shrinkage and swelling lend themselves more readily to quantitative analysis than do the other intrinsic artifacts caused by fixatives. Interesting experiments can be made with the same gelatine-albumin gel as is used in studies of rate of penetration. Acetic acid swells strongly, and indeed this is one of the main reasons why it is included in so many fixative mixtures, for the swelling counteracts the shrinkage caused subsequently by embedding, especially if the swollen cells are fixed in their swollen state by a coagulant fixative. The given figure shows that most of the primary fixatives do not alter the volume of the gel greatly, but ethanol shrinks it strongly.

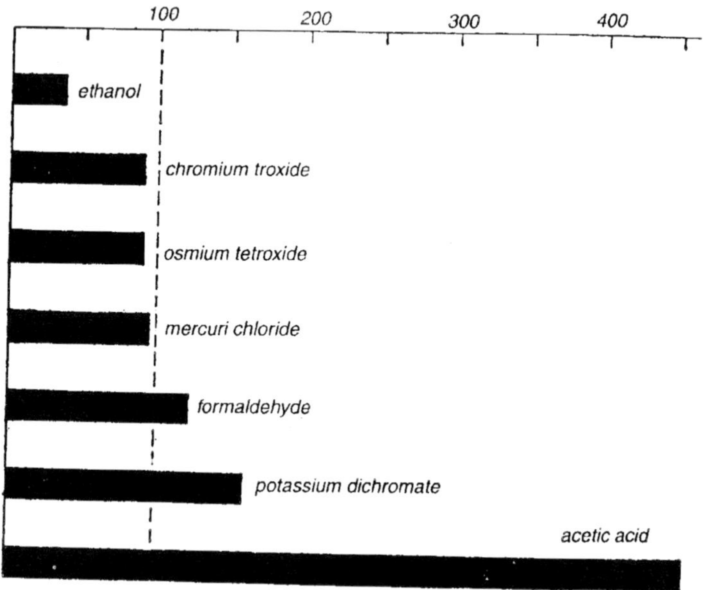

Figure 1.3 : ***Diagram showing the changes in volume undergone by gelatine-albumin gel when acted upon for 18 hours by the selected primary fixatives. The latter were used at their standard concentrations.***

Whole animals or whole organs may be measured before and after fixation, and at various stages in the subsequent processes of microtechnique; or cubes may be cut out of large organs and treated in the same way. The most interesting results, however, are those obtained by the study of single cells. It is best to choose loose, spherical cells for this purpose, since they are simplest from the geometrical point of view. The eggs of echinoids are suitable in this respect. Some results are shown in Table 1.1.

Comparable results are shown by studies of the primary spermatocytes of the snail, *Helix aspersa.* When they have been fixed in formaldehyde solution, embedded in paraffin, sectioned, and mounted in Canada balsam, their volume is only 34% of what it was when they were alive; yet formaldehyde is superior to the six other primary fixatives in this respect. Ethanol is the worst of all: the final volume is only 19% of the original.

TABLE 1.1

The volumes of unfertilized eggs of the echinoid, Arbacia pustulosa, *at various stages of preparation for microscopical study, expressed as percentages of the volumes while alive in sea-water. Each figure is calculated from the mean diameter of 10 to 15 eggs.*

Fixative	***Volume in***			
	fixative, 24 hours	*ethanol, 70%*	*ethanol, abs.*	*xylene*
formaldehyde (in sea-water)	105	70	50	48
ethanol	48	—	—	41

The addition of 'indifferent' or non-fixative salts to fixatives often improves the results obtained. A saline solution may be prepared, having about the same osmotic pressure as the body-fluids of the organism from which tissue is to be taken, and the fixative substance dissolved in this instead of distilled water. It follows that the complete solution is hypertonic. This would be thought likely to cause great shrinkage of the cells, but in fact shrinkage is reduced by the presence of the indifferent salt, and indeed that is precisely why it is used. The way in which it acts is obscure .[21] The use of an indifferent salt is particularly helpful when the fixative is chromium trioxide or formaldehyde.

To make a general study of the artifacts caused by fixation, separate cells are often examined in a suitable saline solution while still alive, and a fixative is then allowed to run under the coverslip. The immediate effects are thus easily observed. The classical work of this kind was done by dark-ground microscopy. The introduction of phase-contrast microscopy resulted in a renewal of interest in work of this kind. The method is valuable but open to two objections. First, it is clearly a test of preservation, not of fixation. Secondly, it shows only how fixatives affect cells that come into immediate contact with them. Many valuable fixatives damage the most superficial cells of a piece of tissue but give good results at a little depth below the surface. When separate cells are examined, they all react like superficial cells, and the fixative may be wrongly condemned.

To overcome these objections, small pieces of suitable tissue may be fixed, embedded, sectioned, dyed, and mounted, and the resulting preparation carefully compared with what can be seen in living cells from the same material. It is best to choose tissues that are readily available but difficult to fix well. The testes of various animals (mammals, urodeles, insects) are suitable. Embedding in paraffin is particularly likely to cause distortion, and this medium should therefore be chosen for a rigorous test. The finished slides of the same material fixed in different ways should be carefully compared by an observer who does not know which is which. There are strong prejudices in favour of particular fixatives, and few cytologists would be able to judge the slides impartially if they knew which fixative had been used in the preparation of each.

2

FIXATIVE

The fixatives that are commonly used without the admixture of other primaries are formaldehyde and osmium tetroxide. Solutions of primaries used in practice in this way may be called *simple fixatives* to distinguish them from *fixative mixtures,* which contain two or more primaries. Simple fixatives usually contain 'indifferent' or non-fixative substances in addition to the primary. Two of the indifferent substances that are used with formaldehyde will be mentioned in the present chapter.

The great majority of practical fixative solutions are mixtures. The formulae for a very large number of these have been published. It is evident, from a study of the papers in which the formulae first appeared, that most of them were not the product of scientific experiment based on knowledge of the properties of their components. On the contrary, they were put together in a hit-or-miss fashion. In several cases the formula was relegated to a foot note, with no indication of any reasons governing the choice or concentration of the ingredients. Some of these empirical fluids gave good results and found favour, others did not. A process of natural selection of almost random variations resulted in the survival, on the whole, of the fittest; though many that are used are superfluous.

To understand the action of fixative mixtures, it is best to make a careful study of a few valuable ones. The principles involved will emerge, and will be found widely applicable. The mixtures chosen for study in this book are those of Clarke, Zenker, Flemming, Helly, and Altmann.

The reader may care to refer to the general remarks on the period of fixation.

FORMALDEHYDE IN SIMPLE FIXATIVES

When used in this way in light-microscopy, formaldehyde is generally dissolved in a solution of an indifferent salt. The latter is used at a concentration that gives the same osmotic pressure as the body-fluids of the organism from which a part is to be taken for fixation, or at a concentration slightly less than this. The following is suitable for the tissues of most vertebrates other than elasmobranchs, and for many terrestrial and fresh-water invertebrates:

Distilled water	83 ml
Sodium chloride, 10% aq.	7 ml
Formalin	10 ml

Keep powdered calcium carbonate in the fluid. Wash out with 50% ethanol.

In this and other formulae, the word 'formalin' means the commercial fluid containing formaldehyde at approximately 40% w/v. In the fixing solutions given here, formaldehyde is therefore used at its standard concentration.

For the tissues of marine invertebrates, one may dilute formalin with 9 times its volume of sea-water. The pH of the fluid is 7.6.

There can be no doubt that non-fixative salts improve fixation by formaldehyde and certain other primary fixatives. Since they are used at concentrations that give about the same osmotic pressure as the body-fluids of the organism, the fixative solution as a whole is hypertonic, for the osmotic pressure of the formaldehyde or other fixative is added to that of the non-fixative salt. One would therefore expect the latter to cause shrinkage: in fact, however, the virtue of non-fixative salts is that they *reduce* shrinkage. It has already been mentioned that this has never been satisfactorily explained. It has been suggested that when fixatives are used without indifferent salts, the cells in the interior of a piece of tissue behave at first—before the fixative substance itself has arrived—as though the piece had been placed in distilled water: that is to say, they swell up and burst. After bursting, they are thought to shrink, and to be fixed subsequently in this shrunken condition. If an indifferent salt is used, it diffuses into the tissue ahead of the fixative, and prevents the initial swelling and bursting. Certain difficulties in accepting this hypothesis as a complete explanation have been discussed rather fully elsewhere.

Formaldehyde is much used in these simple solutions in histochemical studies, especially of lipids. To make sure that phospholipids will be fixed, the tissue may be 'postchromed'. After quite short fixation (6 hours) in the formaldehyde solution, it is transferred (without washing) to a solution of potassium dichromate. A solution of this salt maintained at saturation in an incubator at 37°C is suitable. Treatment for a day or two suffices. The tissue must then be washed in running water overnight. More elaborate methods of postchroming are also available."

For some reason, not yet satisfactorily explained, it is helpful to add sucrose to formaldehyde solutions in electron-microscopical studies. It may be added at 7.5% w/v or there abouts.

OSMIUM TETROXIDE IN SIMPLE FIXATIVES

Osmium tetroxide gives excellent results when unmixed with other primary fixatives, provided that a suitable embedding medium, such as methacrylate is used. Gross shrinkage and distortion occur if tissues fixed in this way are embedded in paraffin.

Osmium tetroxide has been and is enormously used in fixation or electron-microscopical studies. The reader is referred to figure 2 for a diagrammatic representation of the result produced, and for a comparison with the effect produced by fixation of the same kind of cell with formaldehyde.

It is generally believed that osmium tetroxide fixes best for electron-microscopical studies if the solution is buffered with Michaelis's buffer (slightly modified) at about pH 7.4, but there is evidence from several sources that a simple 1% w/v solution in distilled water works as well or better.

We turn now to fixative mixtures.

Clarke

Absolute ethanol	3 vols
Acetic acid (glacial)	1 vol

(The constituents are not ionized and pH has therefore no meaning) Wash out in absolute ethanol.

The formula for this fixative was given by the English neurologist no less than 35 years before Carnoy, the celebrated Belgian cytologist, first published it. In the intervening period it was familiar in microtechnique as *die Clarke'sche Vorschrift*. This is the most ancient of all fixative mixtures commonly used in microtechnique today.

The advantage that may be gained by mixing certain primary fixatives cannot be more vividly illustrated than by a consideration of Clarke's hardy centenarian. By themselves, ethanol and acetic acid are both very bad fixatives, but each compensates neatly for the defects of the other. The shrinkage that ethanol alone would cause is offset by the swelling action of acetic acid; the latter stabilizes nucleoproteins, which are left unfixed by the former; acetic acid fixes neither cytoplasm nor nuclear sap, but both are fixed by ethanol (the former in a rather coarse coagulum). The mixture is useful in routine micro-anatomy and histology, of some value in studies of chromosomes, and applicable also in histochemistry whenever it is desirable to avoid additive fixation and the resultant shift in the iso-electric points of proteins; glycogen is preserved, but not fixed. The fluid is a solvent of many lipids; for this reason, and because it coagulates rather coarsely, it is unsuitable for the study of most cytoplasmic inclusions.

Zenker

Distilled water	100 ml
Mercuric chloride	5 g
Potassium dichromate	2.5 g
Sodium sulphate	1 g

To 20 ml, add 1 ml of glacial acetic acid immediately before use.

pH 2.5.

Pieces of tissue need not be very small.

Short fixation (3 to 6 hours) is best, since certain dyes do not work well when it is prolonged. The reason for this has not been explained. Wash overnight in running water.

If the tissue is to be dehydrated by means of cellosolve, pass it directly from the fixative to 50% cellosolve and from this to a 0.5%w/v solution of iodine in undiluted cellosolve.

Zenker's fluid is chosen as our next example after Clarke because it is one of the best fixatives for use in routine histology and in preliminary work with unknown tissues in biological microtechnique of all kinds. It contains two protein coagulants (mercuric chloride and acidified potassium dichromate) and a substance that opposes shrinkage (acetic acid). No non-coagulant fixative of protein is included in the mixture. Sodium sulphate is an 'indifferent' salt; its effect in this mixture is uncertain.

The ground cytoplasm and certain cytoplasmic inclusions are far better preserved by Zenker than by most routine fixatives. The fine texture of the protein coagulate is probably due to the mercuric chloride, which also ensures easy dyeing. Acidified potassium dichromate favours the action of acid dyes, which give much more brilliant effects than after Clarke.

The only notable defect of Zenker is a tendency to crumple collagen fibres.

Flemming's Strong Fluid

Distilled water	0.8 ml
Chromium trioxide, 5 % aq	0.3 ml
Osmium tetroxide, 2 % aq	0.4 ml
Acetic acid, 20% aq.	0.5 ml or less

As a general rule, use the full concentration of acetic acid.

pH 1.4; the amount of acetic acid used does not affect the pH.

Use pieces 2 mm or less in thickness.

Wash out for several hours in running water or repeated changes.

Since 2 ml of the mixture are sufficient for the fixation of the small pieces of tissue that should be used, it is desirable to follow the formula given above. There is then no wastage of the extremely expensive osmium tetroxide. The fluid made up in this way has the same composition as that which results from the use of Flemming's own formula.

Flemming contains a trio of ingredients that are found over and over again in successful mixtures; namely,

(1) one or more coagulants of protein (in this case chromium trioxide);

(2) a non-coagulant fixative of protein (in this case osmium tetroxide);

(3) acetic acid.

One might suppose that the best results would be given by non-coagulant fixatives in the absence of coagulants; but tissue that has been treated with no other fixative of protein than a non-coagulant does not give ready access to paraffin, and the preparation of good paraffin sections is often difficult. This particularly applies to osmium tetroxide. The spongework produced by coagulants provides spaces into which melted paraffin càn enter. Beyond this, coagulants give the best fixation of chromosomes, and chromium trioxide is pre-eminent in this respect. If, however, no non-coagulant fixative of proteins is included, the spongework tends to be unduly coarse, and cytoplasmic inclusions are distorted mechanically or even destroyed. The coagulants and non-coagulants thus compensate for one another's defects.

The chief effect of acetic acid in these mixtures is the prevention of excessive shrinkage. Tissues fixed in acetic acid alone swell, but are greatly shrunken on subsequent dehydration. If, however, they are stabilized in the swollen condition by the simultaneous action of a fixative for proteins, the shrinkage seen in the final preparation is far less. When acetic acid is used in mixtures at 5% or thereabouts, there is no necessity to add an indifferent salt. Acetic acid also plays a part in the fixation of nucleoprotein, but this is not important when the coagulant is chromium trioxide.

A defect of Flemming is that the constituents penetrate at different speeds. The most external part of the piece of tissue is not capable of being dyed successfully, because osmium tetroxide has exerted too powerful an effect upon it; the most internal part shows the effects of fixation by chromium trioxide and acetic acid alone (apart from the blackening of lipid globules). In the intermediate region, however, fixation is usually excellent. Chromosomes are well fixed, as indeed one would expect; for the famous German cytologist who designed the mixture did more than anyone else to clarify the process of mitosis. The ground cytoplasm preserves much of its homogeneity; various cytoplasmic inclusions, especially the idiozome, are very well shown. Mitochondria are not necessarily destroyed, but as a rule they are not easily shown after fixation by Flemming with full acetic. In the study of these particular cytoplasmic inclusions the acetic acid is commonly much reduced in amount or omitted.

Helly

Distilled water	100 ml
Mercuric chloride	5 g
Potassium dichromate	2.5 g
Sodium sulphate	1 g

To 10 ml add 0.5 ml of formalin (neutralized by powdered calcium carbonate) immediately before use.

pH 3.7.

Fix overnight and wash for several hours in running water, or leave for 24 hours, between fixation and washing, in a saturated aqueous solution of potassium dichromate maintained at 37°C. Dehydrate through grades of ethanol, with iodine treatment.

The Viennese anatomist's fixative is one of the most valuable in the routine study of cytoplasmic inclusions in paraffin sections. It contains a coagulant and a non-coagulant fixative of protein (mercuric chloride and formaldehyde respectively), and also an important fixative of many lipids (unacidified potassium dichromate). It is uncertain whether the indifferent salt (sodium sulphate) plays a useful role in the mixture.

Helly is not nearly so similar to Zenker as the list of its constituents might suggest. Formaldehyde and unacidified potassium dichromate are profoundly different in their effects on tissues from acetic acid and acidified potassium dichromate.

Mercuric chloride and formaldehyde, acting in conjunction, fix ground cytoplasm smoothly, giving just sufficient sponginess to allow easy penetration by paraffin. The constituents penetrate well and it is not necessary to use very small pieces.

Postchroming is useful in studies of mitochondria. The latter are usually well fixed, though those of mammalian liver tend to shorten and round up.

Altmann

Osmium tetroxide, 2 % aq	1 vol.
Potassium dichromate, 5 % aq	1 vol.

pH 4.0.

Use pieces of tissue 2 mm or less in thickness. In studies of mitochondria fix for 24 hours (postchroming is unnecessary).

Wash out overnight in running water.

Altmann was the first serious student of mitochondria, though he did not know them under this name. The distinguished Leipzig cytologist was mistaken about the nature of these cytoplasmic inclusions, which he regarded as 'elementary organisms', living within the cell; but he designed a fixative that is still one of the best for showing them in preparations for optical microscopy.

The fluid differs from the mixtures already described and from the great majority of fixative mixtures used in microtechnique, in containing no coagulant. Ground cytoplasm is therefore fixed very homogeneously. The cytoplasmic inclusions maintain their form, partly because they are not distorted by coagulation of the surrounding ground cytoplasm, partly because both the constituents of the fluid are lipid-fixatives. The dichromate, a stronger oxidizer than osmium tetroxide, prevents excessive blackening of the tissues and gives easy colouring by acid dyes; unfortunately it dissolves nucleoprotein, and the fixative is not adapted to studies of nucleus or chromosomes.

The disadvantage of this fixative is that it does not give ready access to melted paraffin. Embedding often results in shrinkage and distortion, the tissue cracks, and sections are difficult to flatten. When skill or luck overcomes these troubles, excellent preparations result.

Ammonium dichromate may be substituted for potassium dichromate in this solution. The pH of 'NH_4-Altmann' is slightly more acid. The ammonium salt differs curiously from the other. Mitochondria are not well fixed at the surface of the piece of tissue, but in the interior they are better fixed than by potassium dichromate: filamentous ones retain their form, instead of shortening and thickening. For this reason it is best to cut out rather larger pieces of tissue when the ammonium salt is to be used.

The great majority of fixative mixtures fall into one or other of four groups. Fifteen of the most valuable are listed here.

Group A. **Coagulant** + acetic acid. These are primarily fixatives for micro-anatomy and histology. Examples: Clarke, Zenker.

*Group **B.*** **Coagulant** + non-coagulant + **acetic acid.** These are mostly fixatives for detailed histology and general cytology: many are used in the study of chromosomes. Examples: Flemming's strong fluid, Allen's 'B.15', Bouin, Heidenhain's 'Susa',

Hermann (mammalian formula), Sanfelice.

Group C. **Coagulant** + **non-coagulant.** These are routine fixatives for cytoplasmic inclusions. Examples: Helly, Champy, Flemming without acetic, Mann, Zenker without acetic.

Group D. Non-coagulants only. This is a small group, adapted to the study of cytoplasmic inclusions. Examples: Altmann, Regaud.

It is curious that so many valuable fixative mixtures contain one or more coagulants. This is probably connected with the predominance of paraffin as an embedding medium. Group D is likely to come to the fore in the future, as new plastics replace paraffin;

but it remains to be seen whether any non-coagulant is as useful as chromium trioxide in the fixation of chromosomes for optical microscopy.

NON-COAGULANT PRIMARY FIXATIVES

The non-coagulants form rather a heterogeneous group. Formaldehyde and osmium tetroxide resemble one another in being additive non-coagulants that harden protein gels without separating the water from the protein in them, and fix protoplasm without producing microscopical spongeworks. Potassium dichromate and acetic acid are anomalous. They are not fixatives for simple proteins. The former is used chiefly for its action on certain lipids, the latter for its swelling effect, which counteracts the shrinkage caused by other reagents used in fixation and embedding.

Formaldehyde and osmium tetroxide are of outstanding importance in modem microtechnique. The former is much used in histochemical studies, and both are important in electron-microscopy. They will therefore be treated at greater length than the other primary fixatives selected for description in this book.

Formaldehyde

Standard concentration. 4 % w/v aqueous solution.

Formula. HCHO.

Description. Colourless gas; it is very soluble in water as $HO(H_2CO)_nH$. The monomer, $HO(H_2CO)H$, predominates at the standard concentration. The commercial fluid, 'formalin', is approximately a 40 % w/v solution, containing some methanol. Polymers predominate in this, and the high polymer, paraformaldehyde (n = 100 or more), tends to be deposited as a white precipitate.

Ionization. Formaldehyde itself ionizes to a minute degree, with production of a negligible amount of hydronium ions. It is easily oxidized, however, by atmospheric oxygen to formic acid, and the standard solution, prepared by the dilution of commercial formalin with distilled water, has a pH of about 4. If calcium carbonate is present in excess, the pH is 6.4.

Oxidation potential. Formaldehyde can be reduced to methanol, and thus may act as an oxidizing agent. Its oxidation potential, however, is lower than that of any other fixative, so far as is known (0·23 volt).

Reactions with proteins. This subject is particularly well understood, because it has been necessary for industrial chemists to study it carefully in connexion with the so-called 'tanning' of leather. It will therefore be treated here at some length.

One of the most effective ways of studying additive fixation is to prepare polypeptides consisting of a single amino-acid many times repeated, and to find how much of the fixative they take up from solution. It is only necessary to know how much polypeptide is present and the initial and final concentrations of the fixative substance. This method has been very successfully used in the study of fixation by formaldehyde. Although no protein consists of a chain composed entirely of similar links, yet some are made up of very few amino-acids, and these also are useful in the study of fixation.

Polyglycine is an artificial polypeptide consisting of nothing but a group of glycine molecules, joined by peptide bonds. Similarly, polyglutamic acid consists of a group of glutamic acid molecules, joined in the same way. The fibroin of silk consists mostly of alanine and tyrosine. All these substances bind very little formaldehyde.

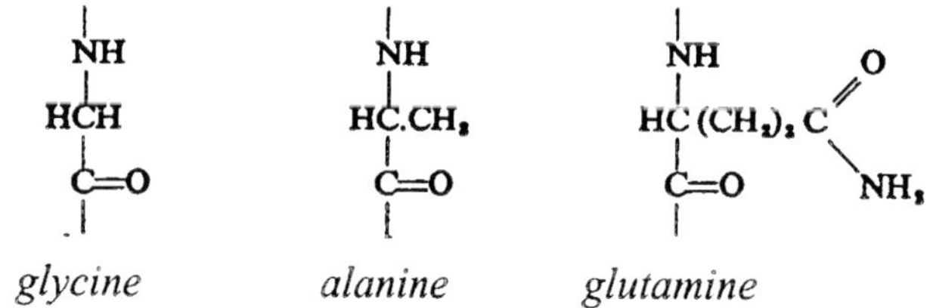

glycine *alanine* *glutamine*

Structural formulae of certain amino-acids, as components o polypeptides and proteins

Polyglutamine is an artificial polypeptide consisting of nothing but a group of glutamine molecules joined by peptide links. This particular polypeptide, in striking contrast to polyglycine, polyglutamic acid, and silk fibroin, binds more formaldehyde than any other macromolecule, so far as is known. A glance at the structural formulae shown here and suggests strongly that the group that binds formaldehyde is the $-NH_2$ of polyglutamine. This conclusion is borne out by experiments made with deaminized proteins; that is to say, with proteins in which the $-NH_2$ groups of polyglutamine and other basic amino-acids have been replaced by $-OH$. These deaminized proteins bind little formaldehyde.

Lysine is the most abundant amino-acid possessing the amino-group, and is thought to be particularly important in fixation by formaldehyde. The reaction might be as follows:

$$-(CH_2)_4NH_2 \quad + \quad H_2CO \quad = \quad -(CH_2)_4NH.H_2COH$$

lysine side-group *formaldehyde*

The $-OH$ group shown in this equation is reactive, and thus a methylene bridge ($-CH_2-$) can be formed between two protein chains, each possessing an amino-group.

$$-NH{-}CH_2{-}NH-$$

A methylene bridge

One can readily imagine two lysine groups on neighbouring protein chains being linked in this way, but analysis shows that there is not a 1:2 relation between the number of molecules of formaldehyde taken up and the number of lysine groups linked to them. It is not probable that two adjacent chains would have many lysine radicles neatly arranged opposite one another, ready to be linked by methylene bridges. Lysine may often in fact be linked to glutamine, or may bind formaldehyde here and there without the formation of a bridge.

Bridge-formation will harden a protein gel, but formaldehyde, as we have seen, is not a coagulant fixative, and it is evident that many of the hydrophil groups retain their relation with water molecules. It is a remarkable experience to transfer a living or fresh jelly-fish from sea-water to formaldehyde solution (preferably dissolved in sea-water) and to see how marvellously its transparency is preserved. This could not happen if the proteins were coagulated. The specimen becomes opaque, however, if after treatment with formaldehyde the water is removed by a dehydrating agent. Similarly, the ground cytoplasm of a cell fixed with formaldehyde loses part of its initial homogeneity when passed through dehydrating agents in the course of paraffin embedding. Formaldehyde does not stabilize the cytoplasm fully against the destructive effects of subsequent dehydration.

Formaldehyde does not coagulate nucleoproteins.

Reactions with nucleic acids. Does not precipitate DNA from solution.

Reactions with lipids. Most lipids are well preserved by formaldehyde, though not necessarily fixed. Triglycerides and cholesterol, for example, are not dissolved. If lipid-solvents are avoided in the subsequent treatment of tissues fixed by formaldehyde, the lipids of adipose tissue remain in their original sites. Certain phospholipids, however, are very slowly dissolved by aqueous solutions of formaldehyde. (Lecithin is not dissolved.) In dissolving, there is a partial separation of glycerophosphoric acid from the nitrogenous base.

Formaldehyde tends to render phospholipids insoluble *in lipid solvents.* It thus acts as a fixative for lecithin. The chemistry of this process has not been fully worked out, but reaction with the nitrogenous bases has been suggested.

There is some evidence that formaldehyde can also react with unsaturated lipids at the double bonds $\left(\begin{matrix} \text{H} & \text{H} \\ -\text{C} & = & \text{C}- \end{matrix}\right)$ on their fatty acid chains. It would appear that in this reaction formaldehyde acts as an oxidizing agent, with production of aldehyde groups in the lipid. It does not appear to be proved that this change confers insolubility in lipid-solvents.

Reactions with carbohydrates. Glycogen often exists in the cell in intimate relation with protein, and formaldehyde fixes the protein in such a way that this carbohydrate is not easily dissolved out by water. Formaldehyde is therefore a valuable component of fixative mixtures intended for use in studies of glycogen.

Rate of penetration. Fast *(K = 3.6).*

Shrinkage or swelling. Swells gelatine-albumin gel somewhat but leaves the volume of liver almost unchanged.

Hardening. Hardens strongly.

Method of washing out. No special washing out is usually necessary, since formaldehyde is very soluble in water and in ethanol.

Effect on the appearance of cells in microscopical preparations. The form of cells is rather well preserved, though there is a tendency for little blebs of cytoplasm to be nipped off from those directly exposed to the fixative (that is, not protected by overlying cells). Cytoplasm is rendered very finely granular. Mitochondria and lipid droplets are generally well preserved. The nucleus remains remarkably lifelike.

In *paraffin* sections there is considerable distortion. Cell-aggregates are separated from one another by wide artificial spaces; cytoplasm is shrunk towards nuclei. Mitochondria are sometimes retained; most lipid globules have been dissolved away. The interphase nucleus retains a nearly lifelike appearance, but mitotic and meiotic chromosomes are not well fixed.

Formaldehyde leaves proteins in a state in which they readily take up basic dyes. This follows from the fact that their acidic groups are unaffected by this fixative. Their basic groups, on the contrary, are to a large extent blocked, and the affinity for acid dyes is therefore greatly reduced. If the protein were allowed to take up all the formaldehyde with which it was capable of combining, all affinity for acid dyes would presumably be lost. In practice, however, this stage is not reached, because the reaction between protein and formaldehyde proceeds slowly. If soluble proteins or polypeptides be allowed to react with excess of formaldehyde for 8 hours at 70° C, they take up about one-half of the total amount that they are capable of binding, and the reaction is still incomplete after 24 hours. The binding of formaldehyde is of course even slower at room temperature.

Formaldehyde is one of the fixatives most commonly used in electron-microscopical studies. Tissues fixed with formaldehyde are very transparent to electrons; and it is necessary to 'stain' the sections with a substance that is able to deflect electrons because it contains heavy atoms (e.g. those of uranium or manganese). Two of the best electron 'stains' for use after formaldehyde fixation are uranium acetate and potassium permanganate.

If one takes into consideration the profound differences between the chemical reactions of formaldehyde and osmium tetroxide, one can scarcely fail to be surprised by the general similarity between electron-micrographs of cells treated with the two fixatives. The reader will find it more profitable to make a careful study of fig. 2 than to read a full description of the resemblances and differences. He will not fail to note that the membranes of the mitochondria and Golgi saccules do not show clearly after formaldehyde fixation, but the internal 'matrix' (or 'contents') of these organelles takes the electron 'stain' and appears dark in micrographs.

Compatibility with other fixatives. Compatible with ethanol, mercuric chloride, and acetic acid. Reduces osmium tetroxide and potassium dichromate slowly, chromium trioxide quickly.

Osmium Tetroxide

Standard concentration. 1 % w/v aqueous solution.

Formula. OsO_4.

Description. Pale yellow molecular (non-ionic) crystals, soluble in water at about 7 % w/W (in carbon tetrachloride at abou 375% w/W). The crystals melt at 41° C. They begin to sublime at a much lower temperature than this; the gas given off is damaging to the epithelium of the eyes, nose, and mouth. This is by far the most expensive of all primary fixatives (1 g costs about £5 10s.).

Ionization. On solution, it takes up a molecule of water to become hydrogen per-perosmate, H_2OsO_5. This ionizes to a minute extent to produce hydronium ions and $HOsO_5$. The substance can, however, scarcely be called an acid. Acetic is a very weak acid, yet its ionization constant is well over 20 million times as great as that of hydrogen per-perosmate.

Osmic acid, H_2OsO_4, is not used in fixation.

Oxidation potential. A moderately strong oxidizer (the 2% solution has o.p. 0.64 volt). On reduction, the brown or blackish hydrated dioxide, $OsO_2.2H_2O$, is left.

Reactions with proteins. Osmium tetroxide gives no coagulum with albumin solution, but on the contrary renders the protein no longer coagulable by ethanol or heat. It sets strong protein solutions into gels, and stabilizes gelatine gels against solution by warm water. It does not coagulate nucleoproteins.

Osmium tetroxide is an additive fixative, but the exact site of its attachment to proteins is not known. It is capable, in certain

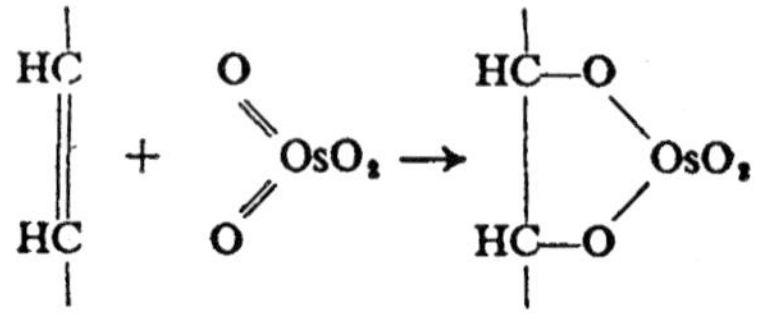

HC=CH + O_2OsO_2 → HC(–O)–CH(–O) OsO_2 (cyclic osmate ester)

Action of osmium tetroxide at the ends of a double bond

circumstances, of simultaneous reaction at both ends of the double bond present in many organic substances. The process is additive.

There are not very many suitable double bonds in proteins, and indeed we have no actual proof that osmium tetroxide reacts with proteins in this way. The evidence does point, however, in this direction.

Most compounds of osmium, other than the tetroxide, are dark or even black, and one can use this fact to find which substances react with the tetroxide and which do not. Of the amino-acids, tryptophane and histidine react strongly, forming dark precipitates. These substances contain double bonds in their five-membered rings. Among the various proteins, the ones that contain a high proportion of tryptophane and/or histidine are particularly reactive with osmium tetroxide. There is a positive correlation between

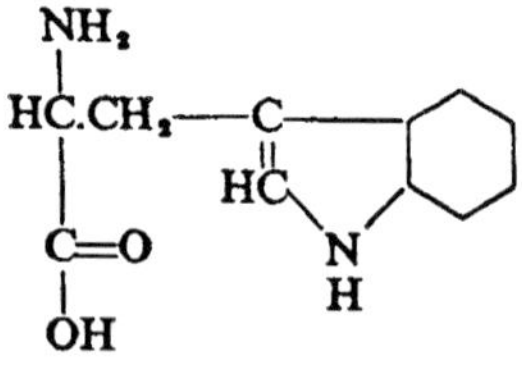

Tryptophane 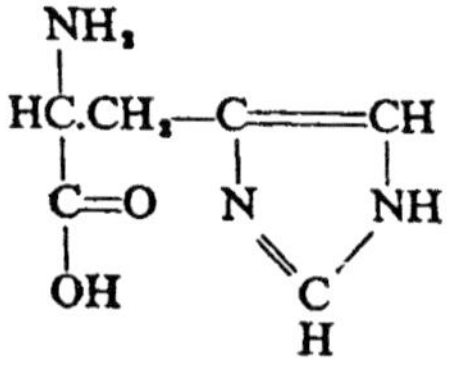 *Histidine*

the tryptophane-content of a protein and the capacity of that protein, in the form of an aqueous sol, to be gelled by osmium tetroxide.

The capacity of osmium tetroxide to gel protein sols and to harden and stabilize protein gels would be more comprehensible if it could be shown that separate protein chains were joined

```
 \         O        /
 HC—O      ||   O—CH
 |     \   ||  /    |
 |        Os        |
 |     /       \    |
 HC—O           O—CH
 /                   \
```

Osmium acting as a bridge between two ring-compounds (The latter are shown only in part.)

together by this substance. Of this we have no positive evidence, but it has been shown that osmium tetroxide can in fact join certain double-bonded ring-compounds together. The double bond is lost in the process. An extra supply of oxygen is required if this reaction is to occur.

Although reactions of this kind are probably concerned in the fixation of proteins by osmium tetroxide, yet there are indications that others also play a part. Tryptophane and histidine are not the only free amino-acids that form a dark precipitate with osmium tetroxide: cysteine behaves in the same way, yet its side-group possesses no double bond.

There is no reason to believe that tryptophane and histidine play any important part in the colouring of proteins by dyes, yet it is characteristic of the proteins of tissues fixed by osmium tetroxide that they show a remarkable lack of affinity for acid dyes. Since these dyes attach themselves to the $-NH_2$ groups of the side-chains of basic amino-acids (especially lysine), there is strong reason to believe that osmium tetroxide somehow blocks these groups. The mechanism of such blocking is, however, unknown. Some proteins that contain a lot of lysine and arginine darken readily with osmium tetroxide, but the protamines, which contain a particularly high proportion of arginine, seem scarcely to darken at room-temperature. It would be unjustifiable to assume that darkening is an invariable concomitant of reaction.

Reactions with nucleic acids. Does not precipitate DNA from solution.

Reactions with lipids. Osmium tetroxide blackens unsaturated lipids of all kinds, but does not attack saturated ones. It is therefore clear that it reacts at the double bonds. Nearly all lipids in organisms are mixtures containing some unsaturated components, and blackening is therefore almost invariable, though it may be very slow.

Osmium tetroxide is soluble in lipids. It may therefore enter a saturated lipid and subsequently be reduced to black osmium dioxide when the tissue is placed in ethanol. It follows that the blackening of a lipid droplet does not necessarily prove that the lipid was unsaturated.

It is an unexplained fact that the conjugated lipids of the tissues, though usually highly unsaturated, are darkened by osmium tetroxide much more slowly than the mixed triglycerides.

If suitable antemedia are used, tissues fixed by osmium tetroxide may be embedded in paraffin .without loss of unsaturated lipids. Benzene and chloroform are particularly suitable antemedia for this purpose.

If unsaturated lipid that has been blackened by osmium tetroxide is bleached by hydrogen peroxide, the lipid is set free and is now soluble once more in lipid-solvents, including benzene and chloroform.

After the fixation of tissues by osmium tetroxide, the unfixed (saturated) component of a lipid globule may be dissolved out during dehydration or embedding. The fixed (unsaturated) lipid is then left in a spherical cavity, which it does not fill. It applies itself to the wall of the cavity, and may be seen in optical section under the microscope as a ring, crescent, or 'cap'. These appearances have often been misunderstood.

Reaction with carbohydrates. There appears to be no reaction in the ordinary circumstances of fixation.

Rate of penetration. This is the only fixative, so far as is known, that does not maintain a constant K-value. It penetrates slowly. The K-value during the first 16 hours is 1.0, but during the period 16 to 144 hours it is only 0.31.

Shrinkage or swelling. Gelatine-albumin gel shrinks very slightly. There are no satisfactory figures for tissues and cells, but there seems to be little change of volume.

Hardening. Leaves tissues rather soft.

Method of washing out. Osmium tetroxide is usually washed out with running water, to prevent subsequent reduction in the tissues by ethanol. In preparations for electron-microscopy, however, tissues are often transferred directly from the fixative to 50% or 70% ethanol.

Effect on the appearance of cells in microscopical preparations. Preserves the structure of the living cell better than any other primary or mixed fixative. It might almost be supposed that the cell was still alive.

There is very poor resistance to the distortion caused by embedding in paraffin. Cell-aggregates are shrunken, so that wide artificial spaces appear; the tissue tends to crack; the cytoplasm contracts round the nuclei; chromosomes (especially in the first meiotic prophase) are poorly fixed.

Acid dyes, as we have seen, scarcely act on tissues fixed by osmium tetroxide. Basic dyes are taken up by both nuclear sap and cytoplasm, and differential dyeing of chromatin is therefore not obtained.

In the early days of electron-microscopy, osmium tetroxide was used almost exclusively in studies of the fine structure of cells, but the introduction of electron 'staining' has made formaldehyde a strong competitor. Most cytologists believe that osmium tetroxide gives a very faithful representation of the fine structure of the cytoplasm and its inclusions, with the possible exception of the nucleus. The slow penetration is not a drawback in electron-microscopical work, since very minute pieces of tissue are used.

A common practice is to fix first with formaldehyde and then 'postfix' with osmium tetroxide before embedding. This process has been used in optical microscopy, and known as 'post-osmication', ever since 1906.

Compatibility with other fixatives. Osmium tetroxide reacts with ethanol and formaldehyde, but is compatible with mercuric chloride, chromium trioxide, and potassium dichromate. These three substances prevent its reduction by daylight.

Potossium Dichromate

Standard concentration. 1.5 % w/v aqueous solution.

Formula. $K_2Cr_2O_7$.

Description. Orange-red crystals, soluble at about 10% w/v in water.

Ionization. The ions produced by the solution of potassium dichromate are the same as those produced by the solution of chromium trioxide, but the proportions are somewhat different. In particular, the amount of hydronium ion is much less, the pH of the standard solution being about 4·1. The chief ion containing chromium is the dichromate, $[Cr_2O_7]^=$, with some hydrogen chromate, $[HCrO_4]^-$.

Oxidation potential. This is a strong oxidizer (o.p. 0.79 volt), but not nearly so strong as chromium trioxide.

Reactions with proteins. Potassium dichromate is a non-coagulant of proteins, including nucleoproteins. It very gradually renders egg-white more viscous and eventually transforms it into a weak gel. It gels histones more powerfully. With these exceptions, it seems doubtful whether potassium dichromate can be regarded as a fixative for proteins, in the circumstances of its ordinary use in microtechnique; but it must be remarked that gelatine gel can be rendered insoluble in warm water by the action of potassium dichromate in bright light.

Since chromium trioxide is a vigorous coagulant, it is evident that proteins react quite differently to the chrome anions according to whether the pH is low or not. The 'critical range' of pH is from 3.4 to 3.8. At more acid pH than 3.4, the chrome anions coagulate protein; above 3.8, they do not. Acidified potassium dichromate acts like chromium trioxide, provided that the pH is below 3.4.

Reactions with nucleic acids. Unacidified potassium dichromate not only does not fix, but actually dissolves DNA.

Reaction with lipids. Potassium dichromate is important in microtechnique chiefly for its fixative effect on certain lipids.

Adipose fat can be rendered insoluble in lipid-solvents by very prolonged treatment with a solution of potassium dichromate. The evidence suggests that the unsaturated lipids are oxidized at the double bonds $\left(\begin{matrix} \text{H} & \text{H} \\ -\text{C}= & \text{C}- \end{matrix}\right)$ the uptake of chromium is not concerned in the process. The reaction is too slow for practical use.

Lipids that have a double bond near the end of the fatty acid chain furthest from the carboxyl group have a special tendency towards polymerization on oxidation; this is accompanied by lessened solubility in lipid-solvents.

Certain lipids are able to take up chromium from solutions of potassium dichromate, and in so doing to lose their solubility in lipid-solvents. This is additive fixation. With the short post-chroming that is usual in microtechnique (for instance, soaking for 24 hours in a 5% solution of potassium dichromate at 37° C), only phospholipids take up the metal. Colour-tests for chromium will therefore reveal the sites of phospholipids in cells. The way in which chromium links itself to the lipid is uncertain. The phospholipids that occur in nature are highly unsaturated, but it is stated that synthetic saturated ones (such as dipalmitoyl lecithin) are able to take up the metal. It seems probable that chromium attaches itself to the phosphate group, though other suggestions have been made.

Reactions with carbohydrates. The unacidified salt is not known to fix any carbohydrate. If acidified, it presumably acts like chromium trioxide.

Rate of penetration. Since proteins are neither coagulated nor gelled in the ordinary period of fixation, the rate of penetration cannot be measured by the method adopted with other fixatives

There is no satisfactory information about the rate of penetration of this substance.

Shrinkage or swelling. Gelatine-albumin gel is swollen by potassium dichromate. Whole livers remain unchanged in volume in a 3% solution, but they are subject to severe subsequent shrinkage by ethanol.

Hardening. Tissues are left very soft.

Method of washing out. To avoid the possibility that insoluble chromic oxide (Cr_2O_3) will be precipitated in the tissues through subsequent reduction by ethanol, it is usual to wash out in running water. This reduction does not occur, however, if light be excluded.

Effect on the appearance of cells in microscopical preparations. The cell maintains its form rather well. Mitochondria are preserved, but filamentous ones may be changed into ovoids. The nuclear sap becomes finely granular. The nucleolus shrinks (presumably through loss of RNA).

In paraffin sections there is a shrinkage apart of cell aggregates, with production of artificial spaces; the cytoplasm, though rather homogeneous, is shrunk round the nucleus; mitochondria are sometimes retained; mitotic and meiotic chromosomes and the heterochromatic segments of the interphase nucleus are unfixed. If the fixative solution is acidified below pH 3.4, the appearance is the same as that given by chromium trioxide.

Both basic and acid dyes act quite strongly. The chromatin, though unfixed, cannot escape through the nuclear membrane; it distributes itself at random within the nucleus. The colouring of it by basic dyes is therefore not informative. This is an unsuitable fixative for studies of the nucleus and chromosomes (unless acidified).

The study by electron-microscopy of cells 'fixed' by unacidified potassium dichromate shows how incomplete this process really is. The cytoplasm is coagulated in such a way as to hold the nucleus in position, but it shows scarcely any of the organelles that are seen after fixation by formaldehyde or osmium tetroxide. The inner nuclear membrane is the one object that is effectively fixed. This membrane prevents the chromatin from escaping into the cytoplasm, but the distribution of the nuclear contents bears no relation to what can be seen in the living cell.

By itself, potassium dichromate is almost useless in microtechnique. If mixed with other fixatives or used for postchroming after fixation, it serves a purpose on account of its action on unsaturated lipids.

Acetic Acid

Standard concentration. 5% v/v aqueous solution.

Formula. $H_3C.COOH$.

Description. A colourless liquid with a pungent smell, miscible with water and ethanol in all proportions. The crystals produced by cooling melt at 16.6°C. The undiluted acid is often called 'glacial', because it is so easily frozen.

Ionization. A very weak acid. The pH at the standard contrition is about 2.3.

Oxidation potential. Can oxidize by being reduced to acetaldehyde (o.p. 0.77 volt).

Reactions with proteins. It neither coagulates nor gels most proteins; non-additive. It extracts histone from the tissues. A thick precipitate is formed when acetic acid is added to a solution of nucleoprotein. This is attributed to the action of the acetate ion in splitting off DNA from protein.

Reactions with nucleic acids. Precipitates DNA from solution.

Reactions with lipids. In writings on fixation it is often said loosely that lipids are dissolved by acetic acid. It is true that some of them (cholesterol and sphingomyelin, for instance) are soluble in glacial acetic acid; but at the concentrations at which it is used in fixation, acetic acid is not a lipid-solvent. It has, however, no fixative effect on lipids.

Reactions with carbohydrates. Neither fixes nor destroys them.

Rate of penetration. In experiments to determine this, it is necessary to substitute gelatine-nucleoprotein gel for gelatinealbumin, for acetic acid leaves no visible mark to indicate its progress into the latter. It penetrates rapidly (K = 2.75).

Shrinkage or swelling. Its capacity to swell protein gels and tissues is the most striking character of acetic acid, and the main reason for its use as a component of fixative mixtures. Its action offsets the shrinkage caused by other substances used in micro-technique. Gelatine-albumin gel expands to between 4 and 5 times its original volume in 18 hours. Acetic acid has much more swelling effect than the mineral acids have. This is partly because the pH is not low enough to cause coagulation; partly, it seems, because the undissociated acid is responsible for some of the swelling. Thus, acetic acid has a swelling (or anti-shrinking) effect even when it is used as a constituent of a non-aqueous mixture, such as Clarke. In aqueous media acids are thought to break the salt-links that connect protein chains; the hydrophil groups thus exposed would draw water into the gel or tissue.

Hardening. Leaves tissues much softer than any other fixative does.

Method of washing out. Since acetic acid is miscible with ethanol in all proportions and has no tendency to produce extrinsic artifacts, no special washing out is necessary.

Effect on the appearance of cells in microscopical preparations. The external form of the cell is fairly well preserved. So, in general, are the cytoplasmic inclusions, but the mitochondria become less visible and may disappear. The nuclear contents are transformed into a coarse network.

In paraffin sections, cell-aggregates tend to be widely separated, with artificial spaces in between. Cytoplasm is sometimes coarsely reticular, sometimes contracted round the nuclei. Mitochondria are absent. The nuclear sap is coarsely reticular; the nucleolus is swollen; metaphase and anaphase chromosomes are rather well fixed; the mitotic spindle appears fibrous.

The acetate ion only produces its characteristic fixation-image on the more acid side of pH 4 or thereabouts. On the less acid side there is no fixation and the tissues macerate.

Cytoplasm takes acid dyes strongly, basic ones rather feebly. The chromatin of interphase nuclei colours feebly with basic dyes, scarcely at all with acid ones. Metaphase and anaphase chromosomes colour strongly with basic dyes.

Compatibility with other fixatives. Compatible with all other fixatives, but potassium dichromate behaves like chromium trioxide if mixed with acetic acid, unless the amount of acid is so small that the pH is on the less acid side of the critical range.

COAGULANT PRIMARY FIXATIVES

Fixatives that stabilize the proteins of the cell quite well may leave much to be desired where cell-aggregates are concerned, for these may be distorted during fixation or embedding, in such a way as to leave artificial spaces. Groups of cells may become separated from other groups or from basement membranes or connective tissue. Although this book is primarily concerned with cytology, yet the needs of the histologist and microscopical anatomist must be kept in mind, and the effects of the selected primary fixatives on cell-aggregates will therefore be briefly noted.

The properties of the selected fixatives are summarized on the pages listed here:

coagulant

ethanol

mercuric chloride

chromium trioxide

non-coagulant

formaldehyde

osmium tetroxide

potassium dichromate

acetic acid

When added at the proper concentration to a solution of albumin, coagulant fixatives separate the protein from the water as a curd or coagulum. This reaction serves to define them. They transform the proteins of the cytoplasm (and often the nuclear sap as well) into a sponge-work. This does not necessarily destroy structure at the microscopical level, since the meshes of the spongework may be very fine; but non-coagulant fixatives are nearly always used in electron-microscopy. The production of a spongework is helpful to the penetration of embedding media, especially paraffin. The coagulant fixatives are much less diverse in their action on cells than the non-coagulants are.

Ethanol (Ethyl Alcohol)

Standard concentration. Undiluted (absolute).

Formula. C_2H_5OH.

Description. A light, colourless fluid, miscible with water in all proportions.

Ionization. Not ionized.

Oxidation potential. Low (95% ethanol, 0·45 volt).

Reactions with proteins. A non-additive or denaturing coagulant of many proteins. It does not coagulate zein, gliadin, or nucleoprotein. Ethanol is a powerful 'unmasking' agent for lipids; that is to say, it sets them free from combination with protein. It acts in this way when used as a fixative and also when used after the tissue has been fixed by a substance that does not itself unmask.[63]

Reactions with nucleic acids. Precipitates but does not fix them.

Reactions with lipids. These are not chemically changed by ethanol, which tends rather to dissolve them than to fix them. This tendency, however, is weaker than is commonly supposed. Tripalmitin and tristearin are insoluble in cold ethanol, and so are several other common lipids of the tissues. Triolein, lecithin, and oleic acid are soluble; other fatty acids and cholesterol slightly so.

Reactions with carbohydrates. Precipitates glycogen without fixing it.

Rate of penetration. Moderate.

Shrinkage or swelling. Shrinks excessively.

Hardening. Hardens excessively.

Method of washing out. Since ethanol is miscible in all proportions with the antemedia used in embedding (and also with water), and since it has no tendency to form an extrinsic artifact, no special washing out is necessary.

Effect on the appearance of cells in microscopical preparations. It produces a coarse coagulum in cytoplasm and nucleus, and destroys mitochondria. Lipid droplets tend to fuse and may dissolve.

In paraffin sections, cell aggregates are shrunken apart from one another; cytoplasm is often piled up against the cell-membrane on the side opposite to that from which the fixative penetrated. Lipid droplets are not present; chromosomes are not distinctly seen.

Most fixatives render tissues more basic or more acidic than they were in the unfixed state, because they react with and block the acidic or basic side-groups of certain amino-acids. As a result, the affinity of the protein for basic and acid dyes is changed. Ethanol is exceptional in this respect. Since it does not attack any of the side-groups, it leaves proteins neither more basic nor more acidic than they were before fixation, and dyes may therefore be used to find whether basic or acidic side-groups pre-dominated in the living tissues. Nucleoprotein remains strongly colourable by basic dyes, but it is not stabilized in position in the cell, and may even eventually be dissolved out. Ethanol is therefore a very poor fixative for nuclei as well as for chromosomes.

Compatibility with other fixatives. Compatible with mercuric chloride, formaldehyde, and acetic acid. Since it tends to be oxidized through aldehyde to acetic acid, it should not ordinarily be mixed with chromium trioxide, potassium dichromate, or osmium tetroxide (though its reaction with the two latter is slow).

Mercuric Chloride

Standard concentration. Saturated aqueous solution.

Formula. $HgCl_2$.

Description. Colourless, needle-shaped crystals. A covalent compound, with much lower melting-point than most salts: sublimes easily. Soluble in water at about 7 % ; readily soluble in ethanol and in benzene. Very poisonous.

Ionization. Ionizes only partially, with hydrolysis, to give

$[HgCl_4]^=$ (particularly abundant in the presence of other chlorides), Hg^{++}, hydronium ions (the pH of the standard solution is about 3·2), and other ions.

Oxidation-potential. Rather a strong oxidizer (o.p. about 0·75 volt).

Reactions with proteins. An additive, coagulant fixative. The reactions depend on the pH at which the fixative is used.

Mercuric chloride is generally used at a pH below the isoelectric point of proteins, and consequently the $-NH_2$ groups of their basic amino-acids will mostly be ionized as $-\overset{+}{N}H_3$. There will therefore be electrostatic attraction between this positively

NH

$HC(CH_2)_4\overset{+}{N}H_3$

C=O

Lysine forming part of a protein chain, in a solution on the acid side of the iso-electric point

charged group and the negatively charged ion, $[HgCl_4]^=$, which could associate itself with such groups in two previously separate protein chains and thus hold them together. The mercury thus adds itself to the protein in the form of an ion. The reaction is promoted by the presence of sodium chloride, since added chlorides increase the amount of the reactive ion.

Mercuric chloride can also react with the cysteine groups of proteins, forming a bridge that connects these groups in previously separate protein chains.

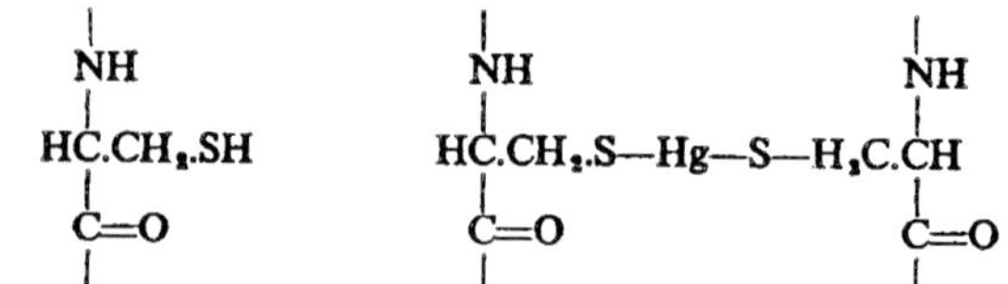

Cysteine as part of a protein chain — *Mercury forming a link between cysteine side groups in two protein chains*

Mercuric chloride does not coagulate nucleoprotein solutions very powerfully: there is flocculation, not formation of a coherent clot. It acts on lipoproteins as an unmasking agent.

Reactions with nucleic acids. Precipitates weakly: *Reactions with lipids.* Mercuric chloride leaves triglycerides untouched. It forms compounds with phospholipids, but their solubilities in lipid-solvents do not appear to have been studied. *Reactions with carbohydrates.* None is described. *Rate of penetration.* Moderate (*K* = 2.2). *Shrinkage or swelling.* Shrinks gelatine-albumin gel and whole livers slightly.

Hardening. Hardens moderately.

Method of washing out. Mercuric chloride leaves black particles, generally a few μ in diameter, in the tissues. These are said to consist of metallic mercury. They can be removed by iodine in alcoholic solution, presumably by formation of mercuric iodide. Iodine colours the tissues, but it may be removed by soaking in 70% ethanol or (much more quickly) by the action of sodium thiosulphate.

$$\underset{\text{Thiosulphate}}{2[S_2O_3]^{=}} + I_2 \rightarrow \underset{\text{Tetrathionate}}{[S_4O_6]^{=}} + \underset{\text{Iodide}}{2I^-}$$

Effect on the appearance of cells in microscopical preparations. While the cell still lies in the fixative, it is better preserved than by any other coagulant fixative. Its shape is well maintained, the ground cytoplasm and nuclear sap are finely coagulated, and mitochondria are not destroyed.

In paraffin sections the cytoplasm is rather badly shrunken and the cells tend to separate from one another; chromosomes are not well fixed ; mitochondria and lipochondria are usually not present.

Proteins are more readily coloured by basic dyes than they are after the action of any other fixative. The –COOH side-groups of their acidic amino-acids are presumably intact in the usual circumstances of fixation. The special affinity for basic dyes must be attributed in part to the coagulation of protein chains in a form that permits the dye ions to penetrate easily between them to reach their sites of action. There is some affinity for acid dyes, and it is therefore clear that the amino-groups of the proteins are not completely blocked.

Electron-micrographs show that mercuric chloride does not

damage the fine structure of cells nearly so much as one would be likely to expect (though it is very far from being a fixative that one would choose for use by itself for electron-microscopical studies). Here and there the ergastoplasm can still be recognized with the accompanying ribosomes, though the sacks ('cisternae') are much flattened, and broken across in many places; the inner and outer nuclear membranes and the mitochondria persist, though the latter show little of their internal structure.

Compatibility with other fixatives. Compatible with any of the selected primary fixatives.

Chromium Trioxide

Standard concentration. 0.5 % w/v aqueous solution.

Formula. CrO_3.

Description. Brownish-red deliquescent crystals, extremely soluble in water (a saturated solution is about 62% w/W).

Ionization. In water it forms chromic acid, H_2CrO_4, which cannot be isolated. A small amount of this remains undissociated in water, while the greater part of it ionizes as dichromate, $[Cr_2O_7]^=$, and the rest mostly as hydrogen chromate, $[HCrO_4]^-$, with sufficient hydronium ions to give a 1 % solution a pH of 1.2.

Oxidation-potential. This is much the strongest oxidizer of all fixatives (o.p. about 1·1 volt), the dichromate ion being readily reduced to the chromic ion, Cr^{+++} or to chromic oxide, Cr_2O_3.

Reactions with proteins. A powerful coagulant of albumin and many other proteins, including nucleoprotein. Chromium trioxide does not fix gelatine gels. It is probably an additive fixative, but the reactions involved are not known. The affinity of proteins for basic dyes is rather low after fixation by chromium trioxide, and this suggests blocking of the –COOH groups of the acidic amino-acids. It has been suggested, however, that acidic proteins are not fixed but simply dissolved away.

Reactions with nucleic acids. DNA is precipitated from solution in an insoluble form.

Reactions with lipids. The fat of adipose tissue can be made insoluble in lipid-solvents by prolonged treatment with chromium trioxide, but the ordinary period of fixation does not suffice. The reactions with lipids have not been studied in detail, but they probably resemble those of potassium dichromate.

Reactions with carbohydrates. Polysaccharides are converted to aldehydes, by oxidation, and will then respond to colour-tests for aldehydes. It is not certain that chromium trioxide actually fixes glycogen, but the aldehyde formed seems less liable to be dissolved out of the tissues by water than glycogen itself is.

Rate of penetration. Slow (K = 1.12).

Shrinkage or swelling. Shrinks gelatine-albumin gel slightly, tissues considerably.

Hardening. Moderate.

Method of washing out. It is usual to wash out in running water. If any free chromium trioxide is left in the tissues, it may subsequently be reduced to green chromic oxide, Cr_2O_3, by ethanol or some other reducer. The oxide is difficult to remove from the tissues, as it is insoluble in ordinary solvents and resistant to acids and other reagents.

Effect on the appearance of cells in microscopical preparations. Ground cytoplasm is rather coarsely coagulated and there may be some distortion of the shape of the cell; the nucleus is on the whole well preserved, the chromosomes excellently.

In paraffin sections, cell-aggregates are quite well fixed; so is the nucleus, though the nuclear sap is changed to a coarse coagulum; the chromosomes and nucleolus are particularly well shown; mitochondria and lipid droplets are not seen.

Cytoplasm has a strong affinity for acid dyes and is therefore easily dyed in a different colour from the nucleoprotein, though the latter has not such a strong affinity for basic dyes as it has after fixation by mercuric chloride or formaldehyde.

Compatibility with other fixatives. It can be mixed with mercuric chloride, osmium tetroxide, or acetic acid. If it is mixed with potassium dichromate, the special properties of the latter substance do not appear. As a general rule it is best not to mix chromium trioxide with reducers, such as ethanol or formaldehyde, with which it reacts. A few useful fixatives, however, contain chromium trioxide and formaldehyde.

FIXATION FOR ELECTRON MICROSCOPY

The cell fine structure of a tissue that is to be examined in the electron microscope must be stabilised by a suitable fixative. Although many fixatives are in use in light microscopy, only a few are used in electron microscopy, the three most important being osmium (OsO_4), glutaraldehyde ($CHO(CH_2)_3CHO$) and potassium permanganate ($KMnO_4$). Which of these is chosen for use depends on the chemistry of the cell structure you wish to examine. Fixatives stabilise cell structure by creating cross-linkages between some of the constituent molecules, thereby providing a framework so that the fine structure does not change through stresses caused by dehydration and embedding. Although it is known that this cross-linking chiefly involves proteins and lipids, the full details of the chemistry of fixation are not well understood. Cell constituents which do not react with fixatives may be washed out; on the whole it is only the macro-molecular skeleton of the cell which is preserved..

The affinities between fixatives and cell constituents are summarised in Table. 2.1.

TABLE 2.1 : REACTION OF FIXATIVE WITH CELL CONSTITUENTS

	Nucleic acids	*Proteins*	*Phospholipids*	*Polysaccharides*	*Unsat. fats*
OsO_4	□	▤	■	□	■
Aldehydes	□	▤	□	□	□
$KMnO_4$	—	■	■	□★	▤

■ strong reaction — destroyed
▤ moderate reaction ★ stains glycogen and lignin or lignin precursors
□ weak/no reaction

Nucleic acids are found primarily as DNA in the nucleus and RNA in the cytoplasm (ribosomes). They do not react well with any of the fixatives, but are usually revealed by the proteins or lipids associated with them. Proteins are found in the form of enzymes, inclusions, secretion granules, etc., and usually need to be stained to increase their contrast even when fixed with OsO_4. Phospholipids are characteristic of most membranes (ER, Golgi). Polysaccharides, deposits of which are sometimes found in chloroplasts, are not fixed by any of the preparations in current use, but owing to their insolubility show up as hazy light areas on electron micrographs. Unsaturated fats react with OsO_4 but saturated fats do not and are leached out.

Osmium is no longer the most popular initial fixative in botanical work, having been largely replaced by glutaraldehyde and glutaraldehyde-osmium double fixation. Potassium permanganate enjoys some popularity in botanical work, but because of its destructive effects on some cell constituents, often leaving only the phospholipid membranes, it is only rarely used in preference to glutaraldehyde or osmium. Most of those who use it exclusively claim that neither osmium nor glutaraldehyde will fix their material adequately.

Fixation times are very variable depending on the material. The correct time for a particular specimen must be determined by experiment, although with glutaraldehyde it appears that time is of little importance. Material is usually well fixed in several hours, but can often be left in the fixative overnight without damage.

Buffers are commonly used at a pH of 7-7.3, although it is thought that a pH of 8 is helpful with highly hydrated material. Veronal-acetate and phosphate are by far the most commonly used buffers.

Preparing Material for Fixation

Preparation of plant material for electron microscopy is easier than that for animals, since the cells do not die so quickly on being removed from the organism. The individual organelles of a cell may, however, react violently to wound stimuli and damage should be kept to a minimum. The primary rule to remember is that fixatives do not penetrate very far into the tissue therefore the tissue must be in the form of thin slices, or preferably in very small cubes.

Place small pieces of leaves, stems, etc. directly in the fixative. The surface layer of cells will have suffered mechanical damage, and should be rejected later, but the inner portion of each cube or slice should be properly fixed. If it is too thick there will be a poorly fixed inner zone which will become apparent in dehydration. The cutting of strips or sheets is useful when a particular orientation must be preserved.

Prepare unicellular algae, bacteria and other finely suspended material by centrifuging them down into a loose pellet, then gently resuspending them in the fixative. After fixation, recentrifuge the suspension. Suspend this pellet again in a very small amount of warm 2% agar, and place it in drops on a slide, or if a large amount, in a thin layer in a petri dish. After setting, cut the agar into tiny blocks which can then be handled as small cubes of tissue in dehydration and embedding.

SPECIAL METHODS

1. In Vivo Fixation

Hallam has developed a method of fixing leaf tissues *in vivo.* The technique can probably be adapted for other tissues. A more complete and extensive fixation of all cells takes place in this way as the cells remain undisturbed until penetrated by the solutions. The procedure is as follows

1. Seal a glass or plastic ring 15 mm wide by 3 mm high onto the upper surface of a leaf with lanoline, taking care that the lanoline does not cover any of the leaf inside the ring.
2. Fill the ring with fixative, using a hypodermic syringe
3. Place a coverslip on top.
4. Change the fixative after 1 hour.
5. After a further ½ hour, cut out the leaf disc inside ring, cut into 1 mm strips in fresh fixative and fix for a further ½ hour.

Preparation of Highly Vacuolated, Senescent, or Damaged Tissue

Mohr & Cocking have devised a method of preparing these difficult plant tissues for the electron microscope.

Fix small pieces of tissue in 0.02M phosphate (Na+) buffered (pH7) 2% glutaraldehyde for 3 hours, then wash 3 times (10 minutes each) in O.1M phosphate buffer. Then post-fix in O.1M phosphate buffered 2% OsO_4 for 3 hours. All fixation is at 4°C.

Dehydration:

30%, 50% 70%, 80% ethanol[1]	15 minutes each
90 % ethanol[1]	30 minutes
1% uranyl acetate in abs. ethanol	30 minutes
abs. ethanol	30 minutes
abs. ethanol	15 minutes
1 part embedding medium,* 1 part abs. ethanol	15 minutes
Embedding medium	15 minutes
Embedding medium	1 hour
Gelatine capsule	44 hours at 55°C.

*Methacrylate-styrene embedding medium:
7 vols. n-butyl methacrylate; stabiliser removed.
+ 3 vols. styrene vinyl[1,2]; stabiliser not removed.

The catalyst-1 % w/v benzoyl peroxide-contains 25 % water; therefore the embedding medium+catalyst is dehydrated using anhydrous calcium sulphate[1]; centrifuge the embedding medium before use.

Mohr & Cocking state that this embedding medium is used because it has a low viscosity, and large block faces are easily sectioned. It is important to stain with uranyl acetate during dehydration as senescing or ruptured cells post-stain very poorly, whereas uranyl acetate in dehydration gives a good stain.

GENERAL FIXATIVES

Osmium Fixatives

These are the most useful and popular single fixatives. Osmium is a fast acting fixative, which is important if fine structure is to be preserved, and it also stains some cell constituents. The colour of a fresh OsO_4 solution is a very pale straw, and if any other colour (red, black, or purple) is observed, discard the sample. There are a number of different buffers in use with osmium, and there seems to be little to choose between them. Remember that all traces of phosphate buffer must be removed as phosphate ions will react with lead stains to form a precipitate of lead phosphate.

Remember that osmium vapour is lethal. Crack the ampoule containing the osmium under buffer in the vessel in which it is to be stored. Use a fume hood with a strong draught when handling osmium.

Palade fixative : This is the original OsO_4 fixative, still much used in electron microscopy.

Veronal buffer	5 ml
0.1N HCl[1]	5 ml
Distilled H_2O	2.5 ml
2 % OsO_4	12.5 ml

The fixative should have a pH of 7.3 and may be kept, in a refrigerator, for a few weeks. Fix for 2 to 4 hours at 0° to 5°C.

Caulfield botanical fixative : This is an adaptation of the Palade formula for use with plant tissue, using sucrose to correct the hypotonicity of the original fixative.

Veronal buffer	1 part
0.1N HCl[1]	1 part
Distilled H_2O	2 parts
2 % OsO_4	4 parts
Sucrose	0.015 g/ml

Fix for 2 to 4 hours at 0-5°C.; pH7.4.

Dalton OsO_4-Chromate fixative : The chromium salts are supposed to cross-link proteins and therefore to aid in stabilising them; the chromium also increases the contrast by acting as a stain. The drawback with this fixative is that the small chromate ions penetrate the tissue more rapidly than the osmium, and thus reach cell organelles in advance of the osmium complex, doing more harm than good. Care must be taken to mix the chrome solution carefully and to check the pH.

$K_2Cr_2O_7$,1 4 % in H_2O, brought

to pH 7.2 with N KOH[1]	1 part
3.4 % NaCl[1]	1 part
2 % OsO_4	2 parts

Fix for 2 to 4 hours at 0-5°C.

Aldehyde Fixatives

Formaldehyde, the first of the aldehyde fixatives to be used, was at first unsuccessful due to damage of the tissue when embedded in methacrylate; the more recent epoxy resin embedding media have solved this problem.

Some botanical material with heavy walls is poorly preserved with OsO_4 as the fixative cannot penetrate properly; formalin followed by OsO_4 is often successful. Fixation in aldehydes, unlike osmium, does not have a destructive effect on enzyme activity, and permits histochemical studies to be made on sections. Where histochemical investigations are intended tissue should not be post-fixed in osmium. Veronal buffers cannot be used with the aldehydes, as they react. Any other buffers are satisfactory.

Formaldehyde fixative:

Phosphate buffer	50 ml
Formalin[1,2,3] (methanol free 40% formaldehyde)	5.5 ml

Fix for 15 minutes to 12 hours, at pH 7.3. The formaldehyde must be methanol-free, as methanol has a destructive effect. If methanol-free formaldehyde is not available, start with paraformaldehyde and make formaldehyde by adding it to slightly alkaline water at 60°C.

Glutaraldehyde fixative : This fixative, introduced in 1962, has produced a considerable change in fixation procedures. It cross-links in the same way as formaldehyde, but is more controlled. It appears to preserve microtubules, particularly in plant tissue, much better than osmium does. The concentration of glutaraldehyde in the fixative appears unimportant; everything from 0.5% to 12% has been used successfully in botanical work, with the 3%-6% range the most popular. Like formaldehyde, this fixative is compatible with all buffers except veronal acetate.

3% Glutaraldehyde fixative

Phosphate buffer	88 ml
25% Glutaraldehyde	12 ml

Store in the cold and use within 24 hours. Fix for 15 minutes to 12 hours at pH 7.3.

Permanganate Fixation

This fixative is chiefly useful for preserving the lipoprotein complexes in cell membranes; most other cellular constituents are destroyed. It should not be used at all for some delicate tissues such as phloem (Johnson). The use of a buffer is unnecessary when fixing botanical material with permanganate. Mollenhauer recommends a 2% to 5% solution of unbuffered $KMnO_4$ at room temperature (22°C.), with fixation times from 2 to 75 minutes.

Double Fixation

Two fixatives, often mutually incompatible, can be used in sequence to obtain better fixation than would be obtained with either alone. The image resulting from double fixation is the most complete yet obtainable.

The most common double fixative is glutaraldehydeosmium. The less toxic glutaraldehyde used first enables the tissue to be handled during fixation. Since glutaraldehyde penetrates faster than OsO_4, cellular activity stops more quickly which results in better preservation of fine structure.

It is most important to remove all traces of the glutaraldehyde from the tissue before post-fixation in osmium. The two react to produce a fine electron-dense precipitate which will ruin the preparation.

Wash in several changes of buffer for two to four hours to remove all the glutaraldehyde. Only cacodylate buffer can be used for both fixatives.

1. Place tissue in buffered 3% glutaraldehyde for 2 hours or longer at room temperature.
2. Wash in the same buffer, two hours with three changes.
3. Transfer to buffered 2% OsO_4; preferably use same buffer as for glutaraldehyde, fix for 2-6 hours.
4. Wash in buffer.
5. Dehydrate and embed.

Buffers

Veronal-acetate:

Sodium veronal	14.7 g
Sodium acetate (hydrated)	9.7 g
Dist. H_2O to make	500 ml

Phosphate (Millonig's constant osmolality):

A. 2.26% NaH_2PO_4[1] H_2O

B. 2.52% NaOH[1]

C. 5.4 % Glucose[1]

D. 41.5 ml A+8.5 ml B (mix just before making buffer)

Buffer: (pH7.3)

45 ml D

5 ml C

(Glucose may be omitted if osmolality not important)

Sodium cacodylate:

A. 0.1 sodium cacodylate[1]

B. 0.1 N HC1[1]

A	**B**	**pH**
100 ml	8.3 ml	7.2
100 ml	5.4 ml	7.4

Dilute to 200 ml with water.

May be used with osmium, formaldehyde or glutaraldehyde.

WASHING AND DEHYDRATION

Specimens must be washed thoroughly after fixation. Any traces of permanganate or osmic acid remaining in the tissue will be reduced during dehydration in ethanol and leave an electron-opaque precipitate throughout the tissue. Wash the tissues in buffer, if the fixative was buffered, otherwise use distilled water.

Remove the fixative from the specimen bottle with a Pasteur pipette and pour in the water or buffer. Use two changes, allowing 30 minutes in each. Again using a Pasteur pipette, replace the second wash with 25% alcohol. Dehydrate the specimens through the series 25%, 50%, 75%, 100% alcohol, allowing 30 minutes in each. Vacuolated plant cells are very much more susceptible to damage during dehydration than animal or fungal tissue and it is important not to skimp any of the dehydration stages.

All the stages from fixation up to the completion of dehydration must be carried out without a break. Specimens can be left overnight once they are in 100% ethanol. Ethanol is immiscible with polyester embedding media, and has a poisoning effect on some epoxy resins, so the tissue must be soaked in acetone before embedding. Give two washes of 30 minutes each in acetone before soaking in the acetone/resin mixture as described in the following chapter. Soaking in acetone is not needed when embedding in methacrylate and can be omitted when using Araldite. It is, however, essential when using Epikote (Epon) or polyester resins such as Vestopal.

Tissues can be dehydrated directly in an acetone series. Prolonged soaking in 100% acetone is advisable for complete dehydration and it may be necessary to use a drying agent to ensure that the acetone is anhydrous. (This is less important with Epikote than with other media. Epikote is partially miscible with water and the pre-soaking in resin mixture will make sure that dehydration is complete.) Long soaking in acetone makes the tissue blocks brittle and they must be handled with care when they are transferred to the capsules for embedding.

When embedding in water-soluble media, the embedding medium itself can be used as the dehydrating agent. Schedules for using glycol methacrylate and Aquon in this way are given in the next chapter.

3

EMBEDDING

In the early days of histology and cytology, sections were cut by hand. Skilled workers could cut ordinary plant tissue admirably, for each cell was held in place by a firm cell-wall. Zoologists were at a disadvantage, *which* they sought to overcome by using fixatives that hardened the tissues. Indeed, their attention was focused on this, rather than on the primary object of fixation; and accordingly they called their fixatives 'hardening agents'. When microtomes came into general use, it was found convenient to embed tissues in suitable media which would hold the cells and intercellular matter in place while the razor slid past them. Animal tissues are seldom sectioned today without having been embedded, and it is therefore unnecessary that fixatives should have a hardening effect.

Embedding media must necessarily be substances that are capable of easy conversion from liquid to solid form. The liquid penetrates the tissues to a greater or lesser extent and is then converted into a solid. We shall see that this conversion may involve hydrogen bonding, covalent linkage, crystallization, or polymerization.

It is sometimes suggested that embedding media may be divided into two groups: those that penetrate the cells, and those that merely surround them. Actually, it is doubtful whether any sharp distinction can be drawn. Some embedding media certainly enter cells, and indeed may penetrate their nuclei; but whether any particular medium penetrates any particular kind of cell or merely surrounds it may depend on the fixative used. Those fixatives that coagulate proteins in the form of a coarse spongework leave the tissues in a state that favours the entry of embed ding media. Those that fix homogeneously, without coagulation, tend to render the internal parts of cells inaccessible to most embedding media. Failure to enter cells by no means necessarily renders an embedding medium useless, for it may enter intercellular spaces and give considerable support during sectioning.

Three substances have been chosen to represent different sorts of embedding media. These three are gelatine, paraffin wax, and butyl methacrylate. They have been chosen partly because they illustrate three very different processes of solidification, partly because they are among the most valuable of all embedding media.

GELATINE

The outstanding advantage of gelatine as an embedding medium is that the tissues need not at any stage be dehydrated. It follows that most lipids, even if unfixed, remain in the tissue and can be coloured by lysochromes. Distortion by dehydration is also avoided. There are, however, several disadvantages in gelatine embedding, as ordinarily practised. The gelatine cannot be removed from the tissue; and since it is capable of taking up both basic and acid dyes, it often tends to obscure the view. The ordinary process cannot be relied on to provide sections thinner than about 5μ, and they do not adhere to form ribbons. It has been claimed, however, that by modifying the procedure one may obtain ribbons of sections only 20 mμ thick, suitable for examination with the electron-microscope.

It is commonly supposed that there is some special virtue in formaldehyde that makes this the fixative of choice when tissues are to be embedded in gelatine, but in fact almost any fixative may be used. It is reasonable as a general rule to avoid those that dissolve lipids.

To prepare tissues for embedding in gelatine, it is only necessary to wash out the fixative, generally with running water. If the fixative was one that fixes gelatine as well as the proteins of the tissues, it is important to get rid of it by thorough washing; for it would fix the gelatine when the latter started to diffuse into the tissue, and thus stop its progress. Among the common primary fixatives, however, only formaldehyde and osmium tetroxide fix gelatine.

Gelatine is a chemically modified form of collagen, an insoluble protein distinguished by its high content of glycine, proline, and hydroxyproline.

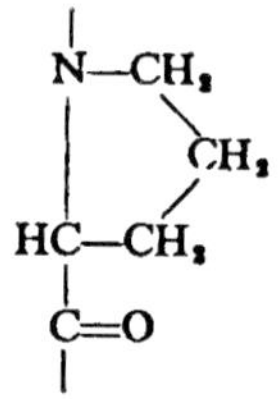

```
 |
 N—CH2
 |    \
 |     CH2
 |    /
HC—CH2
 |
 C=O
 |
```

Proline as part of a protein chain

(*The middle* >CH_2 group is replaced by

>CH.OH in hydroxyproline.)

The collagen used in the preparation of gelatine is derived from the organic matter in the bones of cattle and from the white connective tissue fibres in the skin of cattle and pigs. Bones are first decalcified with hydrochloric acid. The skin or bone is 'limed' for several weeks in an aqueous solution of calcium hydroxide; the soluble proteins are thus dissolved away. After having been washed with water, the material that has resisted solution is kept in clean water at about 60°C for several hours. In this it dissolves. It is filtered, cooled to produce a gel, and dried. The dry material is sold in the form of sheets or powder. The latter is the more convenient form for use in the laboratory.

The transformation that is undergone by collagen during the long treatment with limewater is reflected in the change of the iso-electric point from about pH 7.8 to pH 4.7. This results from the conversion of amide-groups in the asparagine and glutamic acid of the protein chain to the carboxyl groups of aspartic acid and glutamic acid respectively. It is this change that makes gelatine so much more readily colourable by basic dyes than collagen is. Various bonds that tie the protein chains together, such as those between carboxyl in one chain and basic groups in another,

```
 NH                      |
 |       O               NH
 |      //               |       O
HC.CH2.C                 |      //
 |      \               HC.CH2.C
 |       NH2             |      \
C=O                      |       OH
 |                      C=O
                         |
```

Asparagine as part of a protein chain *Aspartic acid as part of a protein chain*

are also broken by the alkaline bath. When the material is warmed, the loosened chains move apart from one another, and are shortened by hydrolytic transverse breakage of the peptide links. Chains varying in molecular weight from about 15,000 to 45,000 are thus set free in association with water as a sol. The molecules are neither fully extended into rods nor compacted into globular form, but each is about 50 times as long as broad.

When the filtered sol is cooled, these molecules attach themselves to one another, probably by hydrogen bonding between peptide groups in previously separate molecules. The bond is very

```
            |
C=O        HCR
 |          |
NH . . . O=C
 |          |
HCR        NH
 |          |
C=O        HCR
 |          |
```

A hydrogen bond (. . .) between two peptide groups in protein chains

susceptible to heat. Over a wide range of concentration, gelatine associates with water to form a solid (gel) below 20°C, a fluid (sol) above 35°C, and a substance showing anomalous viscosity (neither a solid nor a true fluid) at certain temperatures intermediate between 20° and 35°C. There are said to be some free molecules even in the gel.

If gelatine be maintained for a long time at a temperature exceeding 60°C, and especially if it be boiled, the shortening of the protein chain goes so far that the capacity to form a gel on cooling is destroyed. The resulting substance is called metagelatine. Gelatine gels that are intended for embedding should not be melted frequently, since this also has a softening effect.

A piece of fixed tissue that has been washed in water may be transferred directly to a gelatine gel melted at 37°C. There is no reason, either theoretical or practical, why tissue should first be placed in a melted gel of low gelatine content and then transferred to a more concentrated one. A suitable gel may be made by placing 25g of powdered gelatine in 100 ml of distilled water, and leaving this in an incubator at 37°C until the gelatine has dissolved.

If the gel is to be kept in stock, it is necessary to take precautions against the growth of bacteria and moulds. A good disinfectant for the purpose is sodium p-hydroxybenzoate, since this is particularly effective in preventing the growth of the bacterium that liquefies gelatine gels. It should be dissolved at 0.2% in distilled water, and the gel made with this in place of distilled water.

A gelatine gel that has been formed by simple cooling is scarcely hard enough to give the necessary support to embedded tissue, and sections cut from it are inconveniently sticky. It is possible to harden the gel by slow evaporation, but the usual method is to link the molecules more securely together by the action of fixatives. Formaldehyde is suitable. It will be remembered that it acts by forming methylene bridges between protein chains. When the covalent bonds of these bridges have added their effect to that of the hydrogen bonds, the gel is harder; it no longer melts on being warmed, and the tendency to swell in acid solutions is greatly reduced. Sections are still slightly sticky, however. The salts of aluminium are able to fix gelatine. Potassium alum is suitable. It abolishes stickiness and gives hard blocks; but it cannot be used alone, instead of formaldehyde, because sections of gelatine hardened in this way roll up instead of remaining flat. 'Formalum' is a suitable solution for hardening gelatine gels. It is made by diluting commercial formalin with 4 times its volume of a 5% aqueous solution of potassium alum crystals. Gelatine blocks may be preserved indefinitely in formalum.

Formaldehyde solution should be used without the addition of alum if for any reason it is necessary to avoid the mordanting effect of aluminium salts (for instance, in the acid haematein test" for phospholipids; on the subject of mordanting.

Beyond hydrogen bonding and the formation of covalent links, a third hardening process is generally used before sections are cut. The water contained in the fixed gelatine is frozen, usually by allowing carbon dioxide at a low temperature to flow past the block. If, however, the gelatine gel has been hardened by evaporation, little water remains in it and the block can be cut with a glass knife without cooling below room-temperature.

It is possible to attach gelatine sections to glass slides, but it is simpler to dye them (or to treat them with lysochromes) while they are still loose. If loose sections are mounted in an aqueous mounting medium, such as dilute glycerine or Farrants's medium, the tissue will have been kept continuously wet with water from the moment when the cells were still alive.

PARAFFIN

Solid paraffin (so-called paraffin 'wax') is more commonly used than any other embedding medium. It certainly has great advantages. The process of embedding is quick and simple; embedded material may be stored indefinitely in the dry condition; sections may be obtained regularly at all necessary thicknesses from about *2 μ* upwards, if suitable waxes are chosen; and each section adheres automatically to the next as it comes off the microtome knife, so that ribbons are formed. These qualities make paraffin the best embedding medium for most purposes in micro-anatomy and routine histology, and it plays an important part also in studies of chromosomes and other cellular constituents that are not easily distorted or dissolved. The main disadvantages of this medium are the shrinkage that occurs during embedding, and the solution of most lipids. Indeed, the incompleteness of our knowledge of the lipid constituents of cells must be attributed largely to the popularity of paraffin among microtomists.

To prepare tissues for infiltration by melted paraffin wax, it is necessary to dehydrate them thoroughly and soak them in a fluid that is readily miscible with the embedding medium. Ethanol and cellosolve are particularly suitable dehydrating agents.

If all the constituents of the fixative are freely soluble in or miscible with 50 % ethanol, the tissue may be transferred directly to that fluid. If one or more of the constituents of the fixative react with ethanol to produce insoluble material, it is usual to wash the tissue thoroughly in running water after fixation. Chromium trioxide, for instance, tends to be reduced by ethanol to insoluble green chromic oxide, Cr_2O_3; osmium tetroxide to black or brown dioxide (OsO_2 or $OsO_2.2H_2O$). It is worth remarking, however, that tissues fixed in chromium trioxide solution may be transferred directly to 50% ethanol, if 2% v/v of strong sulphuric acid has previously been added to the alcoholic solution.

It is customary to transfer tissues through a series of aqueous alcoholic solutions of higher and higher ethanol content, until absolute ethanol is reached. A suitable series is 50%, 80%, 96%, absolute. Many more grades than these are often used, and special apparatus has been invented to achieve very gradual dehydration. It is doubtful whether any benefit accrues from this. However closely graded the series of ethanol solutions may be, there is always considerable shrinkage at one concentration or another, somewhere between 60% and 90% ethanol. It is a useful experience to judge the appearances of dyed and mounted sections of tissues that have been dehydrated in different ways. A friend may be asked to fix two pieces of the same organ in the same fixative, and to dehydrate them in two different ways: one piece through a closely graded series of ethanols, the other through a coarsely graded series, or even from the washing water directly to absolute ethanol. If he makes a number of slides from each piece of tissue and does not divulge which is which, it will be found difficult or impossible to guess.

If the fixative contained mercuric chloride, iodine should be dissolved at 0.5% w/v in one of the grades of ethanol (for instance, 80%), to prevent the formation of a black deposit.

If the fixative contained no water (Clarke, for instance), the tissue should simply be washed in absolute ethanol.

Although the tissue has now been dehydrated, it cannot be invaded by melted paraffin wax, because ethanol and paraffin are immiscible. It is necessary that the tissue should pass through a fluid that is freely miscible with both. Antemedia' are fluids in which tissues are soaked immediately before they are transferred to embedding media. Some of them have about the same refractive index as dehydrated protein. Since they fill up the spaces in the tissue not occupied by dehydrated protein, they render it more or less optically homogeneous and transparent. They are therefore often called 'clearing' agents, but the capacity to 'clear' is a useless attribute of certain antemedia, and some of the best do not possess it. A large number of more or less harmless fluids are miscible with both ethanol and melted paraffin, and one may choose one's antemedium from among them. Toluene is as good as any for routine use. This light, colourless liquid is so called because it was first obtained by distillation of an oleo-resin

Toluene

exported from Tolu in Colombia; but nowadays it is got from coal tar. Its boiling-point (110°C) is well above the melting-point of paraffin wax. Tissues tend to be distorted (unevenly shrunk) when passed directly from ethanol to toluene, and it is therefore better to pass them instead to a mixture of ethanol and toluene in equal volumes, and from this to pure toluene.

Ethylene glycol mono-ethyl ether ('cellosolve') may be used instead of ethanol as a dehydrating agent. The relationship of this substance to ethylene glycol (the familiar 'anti-freeze') is shown by the structural formulae. It is made by the action of ethanol on ethylene oxide under pressure at high temperature. It is a colourless, odourless liquid boiling at 134°C. It has remarkable powers of dissolving diverse substances, including cellulose nitrate and

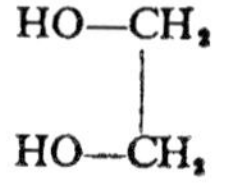

$HO—CH_2$ / $HO—CH_2$	CH_2—O—CH_2 (ring)	$HO—CH_2$ / $C_2H_5O—CH_2$
Ethylene glycol	*Ethylene oxide*	*Cellosolve*

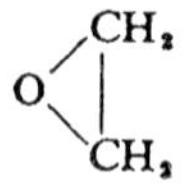

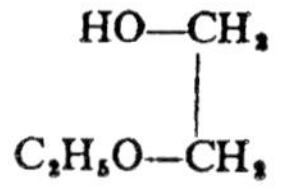

acetate (whence, presumably, its trade name). It is miscible with water and with toluene in all proportions.

Since cellosolve is a less violent dehydrating agent than ethanol, there is no advantage in using a graded series of mixtures of cellosolve with water. Fixed tissues may be transferred from water to a mixture of cellosolve with an equal volume of water, and then to absolute cellosolve; indeed, it is possible to omit the intermediate stage. If the fixative contained mercuric chloride, iodine should be dissolved at 0.5% in the cellosolve; it must subsequently be washed out by the pure solvent. Tissues pass from absolute cellosolve to a mixture of this with an equal volume of toluene, and thence to toluene itself.

Cellosolve causes less shrinkage and hardening than ethanol. Indeed, there are some organs, such as mucous glands, that can scarcely be sectioned in paraffin after dehydration by ethanol, but present no difficulty when cellosolve is used instead. The great solvent power of this substance is, however, a drawback for certain kinds of work. It appears to dissolve some of the constituents of mitochondria, even when appropriate fixatives have been used. In studies of cytoplasmic inclusions it is safer to pass tissues through ethanol, unless it has been shown that the use of cellosolve is not harmful.

Tissues may be transferred directly from toluene to melted paraffin. It would not appear that there is any advantage in soaking them first in a solution of paraffin in the antemedium.

Paraffin waxes are constituents of crude petroleum. They are saturated, long-chain hydrocarbons of the methane series. The commercial products are mixtures of molecules of different molecular weights. The waxes commonly used in microtechnique melt at various temperatures between 52° and 60°C. This suggests that the number of carbon atoms in most of the molecules is less than 30, for $H_3C(CH_2)_{28}CH_3$ melts at 66°C.

Paraffin wax should be maintained at a temperature only a few degrees C above its melting-point. A wax of high meltingpoint (that is, with long molecules) should be chosen if thin sections are required, or if the microtome is to be used in a warm room. There is no advantage in transferring tissues first to a wax of low melting-point and then to one of high. Indeed, the wax of low melting-point would be ousted from the tissue by the other wax very slowly and probably incompletely, for these large molecules replace one another slowly by diffusion.

When the tissue is placed in melted paraffin, the antemedium passes out to mix with it, while the separate paraffin molecules diffuse in to replace them. Great shrinkage of the tissue will occur if the antemedium escapes much more quickly than it can be replaced. If, on the contrary, it escapes too slowly (because it is insufficiently soluble in or miscible with paraffin), replacement is likely to be incomplete. Toluene possesses neither of these defects.

It is best to change the melted paraffin once or twice, in order to get rid of the antemedium that has escaped from the tissue and thus to facilitate the escape of the remainder and its replacement by paraffin.

Tissues are much more fully permeated by paraffin than by gelatine. The embedding medium enters the cells, and indeed there is proof that it sometimes penetrates their nuclei: for paraffin crystals, being birefringent, advertise their presence when sections are examined in polarized light. Penetration takes place readily if the proteins of the cells have been coagulated, but with difficulty if they have been fixed homogeneously. It is partly for this reason, in all probability, that coagulant fixatives have retained their popularity, despite the fact that they necessarily distort the finer structure of the cell. Protoplasm that has been fixed by osmium tetroxide is scarcely porous, and paraffin molecules cannot enter freely.

As a result the embedded tissue, being dry, friable, and not properly supported, often cracks or crumbles during sectioning.

When replacement of the antemedium is complete, the paraffin is hardened by crystallization on cooling. Each long molecule places itself parallel with its neighbours. The molecules are not fully extended, for each forms a zigzag in one plane. Crystallization occurs readily because the molecules, unlike those of certain other waxy constituents of petroleum, have no major branches that would interfere with close packing. Not every molecule has a branch; such branches as occur arise near one end of the molecule, and are usually only one carbon atom long.

The crystals take the form of plates or needles. Each molecule is held in position only by van der Waals forces, and this accounts for the softness of the material and its low melting-point. There is a tendency to a lamellar structure in the crystal, for all the molecular zigzags lie in parallel planes, and each molecule in each monomolecular plane is somewhat more strongly attracted to its neighbours in that plane than to the molecules of other planes. It is presumably the slipping of the planes on one another that gives paraffin waxes their greasy feel.

It is generally believed that paraffin blocks should be cooled quickly, by plunging them into cold water. The wax then hardens in the form of small crystals, and these are supposed to give better cutting qualities than large crystals. An experiment was performed to test this belief. The kidney of a mouse was cut in two. Both the pieces were fixed in Zenker, dehydrated in ethanol, and passed through toluene into melted wax. One piece of wax was cooled suddenly, in the usual way. The cooling of the other was done in stages: first an hour at 45°C, then an hour at 37°C, then further cooling in a warm room (22°C). Both pieces were cut on the same microtome and the sections dyed and mounted. The two blocks cut equally well and the final preparations were indistinguishable.

Sections of tissue that have been thoroughly permeated by paraffin may be freely exposed to the air without risk of damage. This makes it easy to stick them to glass slides, and as a result paraffin sections are very seldom dyed in the loose state. Egg-white is an excellent adhesive for this purpose. It is best to make a stock solution by diluting it with an equal volume of 1 % aqueous sodium chloride solution; any insoluble material is thrown down by centrifuging. In the presence of sodium p-hydroxy-benzoate at 0.2% w/v, this solution resists the attack of bacteria and moulds and may be kept for an indefinite period. For use, the stock solution is diluted with about 50 times its volume of distilled water. The fluid is then spread on a glass slide and the section floated on it. The slide is warmed on a hot plate sufficiently to soften but not melt the paraffin, and any folds in the section flatten out. The water is then drained off as far as possible and the slide left on the hot plate to dry. When it has dried, the minute quantity of albumen between the section and the glass suffices to stick them firmly together.

Glycerine is often mixed with egg-white in the preparation of the adhesive. It serves no purpose.

To remove the paraffin from the dried slide, it is only necessary to soak it for a few minutes in xylene. The slide is then passed through absolute ethanol to 90% and then 70%. If the dye that is to be used is dissolved in a weak ethanol solution, the slide may be transferred to it from 70% ethanol; if the dye is dissolved in water, it is best to rinse in distilled water first.

BUTYL METHACRYLATE

Butyl methacrylate was introduced into microtechnique as an embedding medium in 1949. Other plastics, especially those known by the trade-names of 'araldite', 'epon', and 'vestopal', have to some extent replaced the methacrylates, because they are thought to distort the tissues even less. Unfortunately there is secrecy about their exact composition and about the chemistry of the hardening process, and a description of them is therefore not well adapted to a scientific book such as this, in which the mention of rule-of-thumb methods would be out of place. The general principles underlying the use of plastics as embedding agents in microtechnique are well exemplified by butyl methacrylate.

The great advantage of this and the other embedding media mentioned in the preceding paragraph is that they cause little distortion and permit sections to be cut at any thickness from a few μ down to 50mμ or even less (though for very thin sections a mixture of butyl and methyl methacrylates is used). If tissues are suitably fixed, sectioned in butyl methacrylate at about 3μ and mounted in an appropriate medium, the appearance of the cells is astonishingly lifelike. Methacrylate is seldom used when sections are to be dyed. It finds its chief application in the preparation of very thin sections for electron-microscopy. The main disadvantage of methacrylate is that the process of embedding is rather complicated.

Very small pieces of tissue (often less than 1 cubic mm) are used. After fixation the tissue may be washed with distilled water (though many workers use weak ethanol solutions), and must then be dehydrated. Graded ethanols are used for this purpose. It has not been found necessary, even for the most delicate work, to use a closely graded series of ethanols. The series recommended for use in paraffin embedding (50%, 80%, 96%, absolute) is suitable. No special antemedium is required. The tissue may be transferred to a mixture of equal volumes of absolute ethanol and methacrylate, or directly to methacrylate itself.

The embedding medium to which the tissue is transferred is n-butyl methacrylate, in monomeric form. This is a colourless, mobile liquid, boiling at about 163°C. It has a distinctive, rather unpleasant smell. The vapour is somewhat toxic and it is desirable to increase the ventilation if the smell is noticeable. Butyl methacrylate is miscible with absolute ethanol and with lipid-solvents such as carbon tetrachloride, acetone, and toluene. It is insoluble in water, on which it floats. It is highly inflammable.

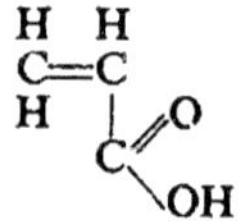

```
H  H              H  H
C==C             —C—C—
H  |  //O         H  |  //O
   C                 C
    \OH               \OH
```

Acrylic acid 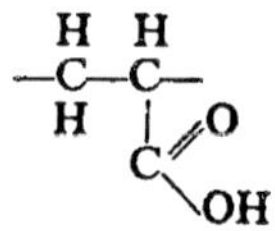 *Acrylic acid as a segmer in a polymer*

The acrylic acids form a series similar to the fatty acids, but two of the carbon atoms in the chain are joined by a double bond. In the strict sense the series includes only those acids in which the double bond is in the same position, in relation to the carboxyl group, as in acrylic acid itself: that is to say, just beyond the next carbon atom from the carboxyl carbon. The aldehyde corresponding to acrylic acid is acrolein, and from this excessively pungent substance the acid derives its name.

From our point of view, the important character of the acrylic acids is their strong tendency to polymerize, and in doing so to become solids possessing special characters. Polymerization occurs by the addition of one molecule to another, the double bond becoming single in the process. There is no condensation: each repeated unit or 'segmer' consists of the same atoms as the monomer. The substance formed is thus an addition-polymer.

The hydrogen of the carboxyl group of the acrylic acids can be replaced by methyl, ethyl, or other alkyl groups, without loss of the tendency of the molecule to polymerize. The esters thus formed are of greater practical value than the simple monomers. Reaction (in a roundabout way) with normal butyl alcohol, $HO(CH_2)_3CH_3$ gives n-butyl acrylate. This, when polymerized, is a very soft, rubbery substance, not adapted to the purposes of microtomy. If, however, a methyl group be substituted for the hydrogen attached to the next carbon to that of the carboxyl group, a marked change occurs in the physical consistency of the

```
                                         CH3
 H H                                   H |
—C—C—                                 —C—C—
 H | //O                               H | //O
   C                                     C
    \O(CH2)3CH3                           \O(CH2)3CH3
```

A segmer of n-butyl acrylate polymer

A segmer of n-butyl methacrylate polymer polymer.

We now have, in n-butyl methacrylate polymer, a hard, colourless, transparent solid, admirably adapted to our needs. All the methacrylates are harder than the corresponding acrylates, but the length of the esterifying alcohol also affects the hardness. The shorter the alcohol, on the whole, the harder the polymer. Methyl methacrylate—better known under its commercial name of perspex—is the hardest of all. It is often used as an embedding-medium for electron-microscopy, usually with the admixture of butyl methacrylate. The latter by itself, however, gives good sections for electron-microscopy as well as for optical microscopy.

If the segmers of butyl methacrylate were only capable of linking together in the way previously mentioned, the substance produced would resemble a fibre rather than a resin. In fact, the long chains branch and link in such a way as to form a threedimensional spongework devoid of any particular orientation. The reactions involved in this process are mentioned below. It is this submicroscopic structure of the methacrylates that fits them so well for use as embedding media, capable of being sectioned equally well in any direction.

At room-temperature the monomer very slowly undergoes spontaneous polymerization. An inhibitor must be added to enable the methacrylate to be stored in monomeric form. Quinone is suitable. It is stated to react with chain-forming radicles to produce stable compounds." Manufacturers add a small proportion of hydroquinone to methacrylate, and this becomes converted to

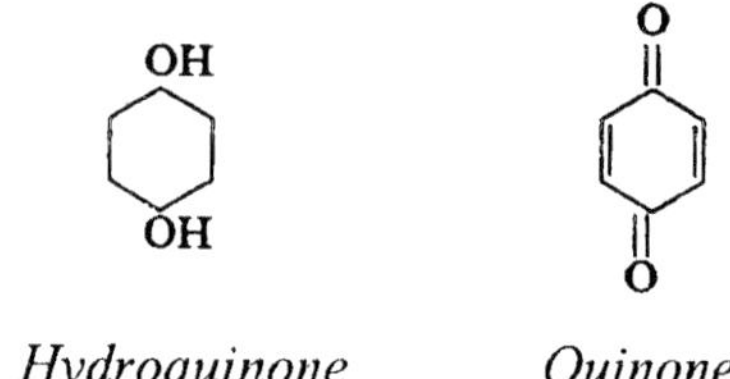

Hydroquinone *Quinone*

quinone by spontaneous oxidation. When the methacrylate is about to be used, the inhibitor is removed by shaking the fluid with an aqueous solution of sodium hydroxide. The water, coloured by the inhibitor, separates from the methacrylate under the influence of gravity. The monomer is washed free from sodium hydroxide by being shaken several times with distilled water; the latter is then removed by the addition of anhydrous sodium sulphate. Although uninhibited, the material will remain unpolymerized for many weeks if kept in a refrigerator maintained near 0°C.

The methacrylate is now ready to receive the tissue. The molecules being fairly small, thorough penetration takes place quite easily. The fluid is changed a few times to get rid of the ethanol. It would then gradually harden by polymerization, if left at room-temperature, but the process is too slow for practical use. It is necessary to add an activator or so-called catalyst and to warm to about 60°C. The name 'catalyst' is inaccurate, for, as we shall see, the substance in question forms part (though a very small part) of the hardened material.

The extremely active substance that promotes polymerization is the free phenyl radicle, H_5C_6. It will be remembered that free radicles differ from the vast majority of chemical compounds in possessing an uneven number of electrons in the sum-total of their constituent atoms. In the formulae shown here, it will be noticed

```
         H                              
         ..                             .
     .·. C ·.                       .·. C ·.
H : C ·     · C : H           H : C ·     ·· C : H
    ..       ....                 ..       ....
H : C ·:   .  C : H           H : C  ·:  ·. C : H
     ·· C ·.                        ·· C
        ..                              ..
        H                               H
```

Diagram of the structure of benzene, showing the valency-electrons

Comparable diagram of the structure of the free phenyl radicle

that 8 electrons of the outer shell are associated with each carbon atom in benzene, but only 7 with one of the carbon atoms in the free phenyl radicle. The formula for the latter is conveniently written $H_5C_6\bullet$ to remind the reader of the existence of the unpaired electron.

Benzoyl peroxide, $(C_6H_5CO)_2O_2$, is used to produce the free phenyl radicle. This colourless, crystalline solid may be regarded as hydrogen peroxide in which each of the hydrogen atoms has been replaced by benzoyl, $C_6H_5CO^-$. It is an explosive, and care should be taken not to grind it with metal instruments; for storage it is damped with water. The tissue is transferred from the unmixed monomer to a weak, freshly prepared solution of benzoyl peroxide in the monomer. It is left in this at room-temperature until the peroxide has had time to penetrate the tissue before active polymerization begins. Since the peroxide was damped, it is necessary once more to dry the methacrylate with anhydrous sodium sulphate. The temperature is then increased to about 60°C and the hardening starts.

Each molecule of benzoyl peroxide loses two of carbon dioxide, and two phenyl radicles are set free. Each phenyl radicle reacts with a monomer molecule, forming a covalent bond with the carbon of the $\overset{H}{\underset{H}{C}}=$ group. The compound necessarily has an uneven number of electrons, but the odd electron now appears in a new

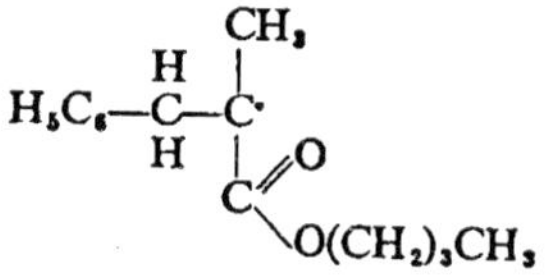

$$H_5C_6-\overset{H}{\underset{H}{C}}-\overset{CH_3}{\overset{|}{C}}\cdot \quad \text{with } \overset{|}{C}\begin{matrix}\nearrow O \\ \searrow O(CH_2)_3CH_3\end{matrix}$$

A phenyl radicle combined with butyl methacrylate
Note the unpaired electron.

place, in the middle of the methacrylate. Meanwhile the double bond has become single. The end of the molecule provided with the odd electron has become as active as the phenyl radicle was, and in exactly the same way. It combines with another monomer, and an unpaired electron is thus produced in the middle of the latter. So the process goes on, and a high polymer is built up. The whole of it consists of methacrylate monomers, except one extremity of the chain.

This process of polymerization leads to the production of very long molecules, but it does not continue indefinitely. The growing end of the chain may come up against another growing end, or against a free radicle; in either case combination occurs, and the chain becomes 'dead'.

The process that has been described would lead only to the formation of molecules having the form of long, thin threads. In fact, however, branching and net-formation occur. Branching takes place in various ways in addition-polymers but the following is probably what happens in methacrylates. A carbon atom, other than a middle one, becomes active. The atom is thought to be one of those in the butyl group. This carbon atom lets go of one of its hydrogens, which departs with its single electron and attaches itself to the growing point of another molecule (which is thus terminated and becomes 'dead'). The carbon atom now has an uneven number of electrons, and therefore acts as a new growing point, to which other monomers add themselves one after another. The branch thus formed may itself be terminated and 'die' as a

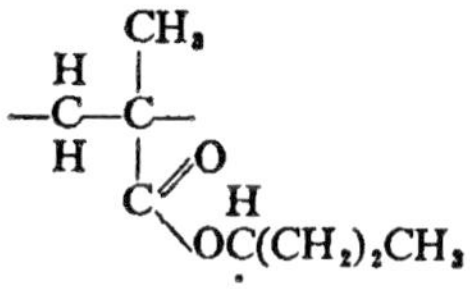

```
      CH3
  H   |
 —C———C—
  H   |  O
      C//   H
       \OC(CH2)2CH3
         •
```

A segmer of butyl methacrylate, with an active carbon (C) in the butyl radicle

loose end; but if it meets another growing branch, the two may fuse and thus accomplish a union between two branched molecules that were previously separate. Since the branches are not all formed in the same plane, a three-dimensional meshwork results.

Tissues shrink slightly in butyl methacrylate while the latter is still in monomeric form. Polymerization involves further shrinkage, since the methacrylate itself becomes reduced in volume. The various methacrylate esters differ from one another in this respect. The longer the molecule of the esterifying alcohol, the less the reduction in volume on polymerizing. Butyl methacrylate contracts to about 85% of its former volume, methyl methacrylate to about 79%. The butyl ester is therefore preferable unless a very hard block is required. Measurements of the volume of the embedded tissue" suggest that this shrinks more than the methacrylate. Shrinkage is less than with paraffin, and since it is nearly equal in all directions, there is little distortion.

Blocks containing tissue can be stored indefinitely in the dry condition. To obtain sections from 8 μ to 1 μ thick, it is best to soak the block overnight in 70% ethanol and to cut on a sliding microtome with an oblique knife flooded with 70% ethanol. The back of the knife should be raised rather higher above the level of the cutting edge than is usual when cutting paraffin. Sections may be stored in 70% ethanol.

Thin sections for electron microscopy are usually cut with knives made by breaking a piece of plate glass along a line at an angle of 45° to one of its edges. The cutting edge is the one at the intersection of the two surfaces that subtend an angle of 45° to each other.

Methacrylate sections may be stuck to glass slides for study by phase-contrast microscopy in the same way as paraffin sections. They are brought from 70% ethanol to water, and floated on very dilute albumen solution on a glass slide. To remove the embedding medium after the section has been attached, it is only necessary to put the slide in ethyl acetate or toluene, which dissolves the polymer. The slide is then passed through the usual series of ethanols to distilled water, and mounted in diluted glycerol.

EMBEDDING FOR ELECTRON MICROSCOPY

Although methacrylate was for many years the only embedding medium used, several resins are now in use. The most generally useful are the newer cross-linked plastics, e.g. Araldite and Epikote.

The specimen must be fully infiltrated by the medium. This is particularly important with viscous media like Araldite and Epikote. A longer time must be allowed for tissue penetration by these media than for thinner plastics like methacrylate.

After dehydration, place the specimens in a 50/50 resin/ acetone mixture and leave uncovered overnight so that the acetone evaporates and the resin slowly increases to full concentration. Then transfer the specimen to a pure resin mixture and allow to stand for a day at room temperature so that full infiltration by the plastic takes place. For particularly difficult tissues, place the specimen in pure acetone in a covered container, to which resin is added slowly over a day to make a 50/50 mixture. Then uncover the container and allow it to stand at room temperature for up to 6 days to bring about complete penetration of the resin. Transfer the specimen to fresh resin and leave for a further day before embedding. Tissue blocks must be embedded in fresh resin. With very difficult specimens raise the temperature up to 90°C. for 20 minutes to aid infiltration.

Embed specimens in size 00 gelatin capsules, which are large enough to hold any tissue specimen and will fit easily into microtome chucks. Hold the capsules upright for polymerization in a box lid in which holes have been punched, or place in wells drilled in ½ in. thick metal slabs. Do not use wooden slabs as uneven polymerization occurs due to poor heat distribution. Specimens should be placed in the top of capsules filled with embedding medium; further infiltration occurs as the specimens sink to the capsule bottom in the oven. If it is necessary (as with very small specimens) to prevent the tissue from dropping to the capsule bottom, polymerize a little resin in the capsule first.

Orientation of the tissue is often necessary. Rods and sheets of tissue will lie with their long axes nearly parallel to that of the capsule, but blocks and small pieces may lie randomly. If a particular orientation is wanted, embed first in a thin layer of embedding medium in a polypropylene beaker or aluminium foil boat; when polymerization is complete remove the hardened block from the container and cut out the tissue with a saw in shapes suitable for mounting in the microtome chuck or for re-embedding in standard capsules.

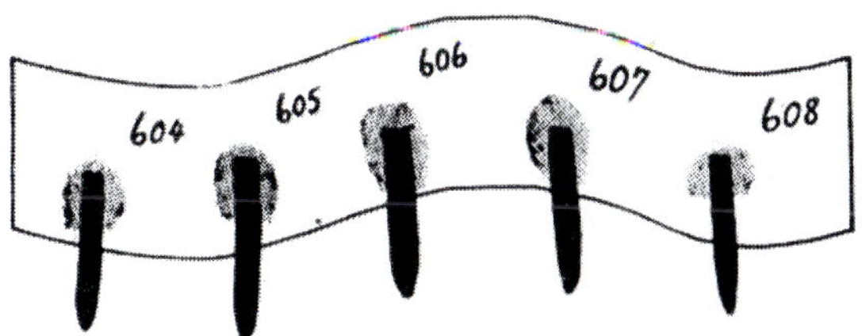

Figure 3.1

The following method is useful when transverse sections of root, shoot or stem tissue are required. Glue the freshlyharvested segments of plant to a strip of stiff paper with Durofix or similar glue unaffected by solvents. Write the block number beforehand on the back of the paper. Transfer the whole unit through fixation, staining, dehydration and soaking and finally cut up into single pieces for the embedding stage. Push the ring of paper and the tissue down into the capsule in stages over about two hours. The tissue will then polymerize in the totally uncontaminated resin at the bottom of the capsule.

Label specimens by placing a small piece of paper with the serial number of the specimen written in pencil in each capsule; the hardened plastic is sufficiently clear to allow the number to be read, and it cannot be lost or rubbed off. Do not use ball-point ink as it is soluble in the resin.

Use disposable polypropylene beakers for mixing the resin.

METHACRYLATE

The oldest of the embedding media, methacrylate, cannot be recommended for general use because of its many defects, the more serious of which include possible tissue damage during polymerization and a tendency for the medium to sublime under the electron beam in the microscope. This evaporation improves specimen contrast, but allows fine detail to collapse. Methacrylate does, however, have some uses in that much larger block faces can be successfully sectioned than with epoxy resins and sections may be more easily stained. Also, the relative thinness of methacrylate

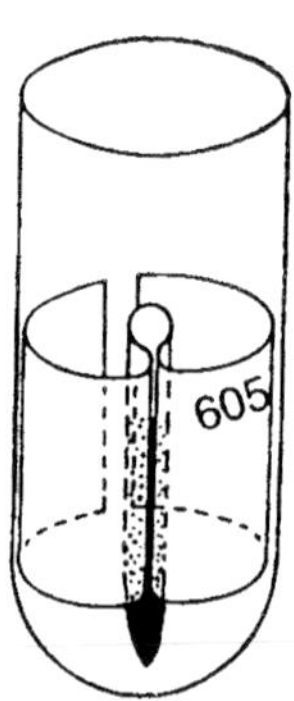

Figure 3.2

monomer enables it to infiltrate tissues which the more viscous epoxy resins may have trouble penetrating. When using methacrylate, encourage rapid polymerization. Raise the concentration of catalyst, increase the temperature to 80°C and introduce a nitrogen atmosphere if possible; oxygen is an inhibitor.

Preparation of methacrylate embedding medium

Methacrylate monomer is sold with hydroquinone added to prevent polymerization. The traditional method involves the removal of this inhibitor by a rather laborious process which can be a source of trouble. As the presence of inhibitor influences only the rate of polymerization and not the properties of the final block, it is not necessary to remove it. Preparation of the medium is as follows:

n-butyl methacrylate	200 ml
methyl methacrylate (more increases hardness)	50 ml
benzoyl peroxide (catalyst)	5 ml

Capsules filled with the above mixture should be placed in an oven at 50°C, with the capsule lids on to prevent evaporation of the monomer.

EPOXY RESINS

Araldite and Epikote are, at present, the epoxy resins commonly used for embedding. Embedded specimens are free from polymerization damage and the plastic does not sublime in the microscope, and so provides continual support and preservation of fine detail.

The resins may be carcinogenic and the accelerator causes severe dermatitis in some individuals. Disposable polythene gloves must always be worn when handling the components. The polymerized plastic is not hazardous. As the resin does not sublime, some sort of staining is almost always necessary to provide contrast.

Preparation of epoxy resins

Araldite:

Araldite M	20 ml
Resin 964B	20 ml

Accelerator DYO64	0.6 ml
Dibutyl phthalate	2 ml (plasticiser)

Mixing of the components must be thorough. It should be carried out in a fume cupboard and a constant stirring and turning motion should be used for 5-10 minutes. Mechanical stirrers are not satisfactory. Hardness of the block is controlled by the plasticiser dibutyl phthalate; less for a harder resin, more for a softer block. Harden in the oven at 60°C for 48 hours. The complete mixture will keep in a desiccator in a deep-freeze for several months.

Epikote:

Epikote 812 (Epon 812)	162 ml
DDSA dodecenyl succinic anhydride	100 ml (softener)
MNA methyl nadic anhydride	89 ml (hardener)
DMP-30 2,4,6-tri(dimethylaminomethyl) phenol	3-4 drops

The mixture must be stirred as for Araldite. The hardness of the final block is controlled by adjusting the relative amounts of softener (DDSA) and hardener (MNA) in the mixture. Heat the mixture to 60°C for 48 hours to complete polymerization. The mixture may be stored in deep-freeze as for Araldite for several months.

ERL-4206

This is a low-viscosity epoxy resin not yet tested in this country, but available in the United States. It has been recommended by Spurr (1969) for use with a wide range of plant and animal tissues and particularly those which are normally difficult to embed such as hard, lignified cell walls and highly vacuolated parenchymatous tissues of ripe fruits.

Its viscosity is so low that, unlike the other epoxy resins, it can be mixed with its other components simply by shaking.

VESTOPAL W

This is a polyester embedding medium with a fine grain. It is quite hard and only small sections are easily cut. Apart from this Vestopal sections well. It stains more easily than Araldite, which can be of advantage with specimens of low intrinsic contrast. It is not miscible with alcohol, so that acetone must be used for dehydration, ending with anhydrous acetone.

Vestopal W embedding medium

Vestopal W[2,3]+ 1 % tertiary butyl perbenzoate (initiator) + 1% cobalt naphthenate[2,3] (activator).

The initiator and activator must not be mixed directly together; the combination is explosive. The complete mixture may be stored in a refrigerator for several months. The activator and initiator do not keep well and must be stored under refrigeration. Even then they will keep only for several months. The Vestopal W medium must be stored in glass as it affects polystyrene containers.

WATER-SOLUBLE RESINS

These embedding media are particularly useful when it is not desirable to dehydrate with organic liquids, or when histochemical tests are to be carried out on the cut sections.

Aquon:

Developed by Gibbons this is not available commercially. It is an extract of Epikote, and the extraction procedure is fairly simple, yielding 30% of the original volume of Epikote. The prepared resin is stable for a year or more if kept stoppered and dry (do not use $CaCl_2$ as a drying agent as it is reactive); the 80 % form is stable for a very short time only. Tissue preservation with Aquon is good, and the block sections well.

Preparation of Aquon

1. Shake one volume of Epikote with two volumes of distilled water.
2. Centrifuge suspension into two layers, until upper layer clears.
3. Remove upper layer, add anhydrous sodium sulphate to excess (until some remains after shaking).
4. Stand until two layers are formed.
5. Remove upper layer; centrifuge at 0°C until clear.

6. Pour off supernatant (an 80% resin solution).
7. Place resin in a shallow tray with magnesium perchlorate[1] in a vacuum desiccator to remove water.

Embedding medium

Aquon resin	10 ml
dodecenyl succinic anhydride (DDSA) (hardener)	25 ml
benzyl dimethylamine (accelerator)	0.35 ml

Dehydration:.

5% Aquon/water	20 minutes	
10%	20 minutes	
20%	20 minutes	
30%	30 minutes	4°C
40%	5 minutes	
50%	5 minutes	
60%	10 minutes	
70%	10 minutes	
80% Aquon /water	30 minutes	
90%	60 minutes	4°C
Pure Aquon	60 minutes	
Pure Aquon	60 minutes	
50/50 Aquon/complete embedding medium	2 hours	room temp.
Complete embedding medium	4 hours	

Place the specimen in fresh resin and harden at 60°C for four days. Store the hardened resin in a desiccator as it is hygroscopic.

Glycol methacrylate:

GMA is a water-soluble embedding resin of considerable value. It allows many compounds to survive with little denaturation and it leaves proteins and nucleic acids susceptible to enzymic digestion in the sections after aldehyde fixation. GMA is commercially available and is fully miscible with water, ethanol[1] and ether. Specimens are dehydrated by passing through a graded series of resin and water as for Aquon. Sections tend to swell after cutting if in contact with water. Use a 0.5M aqueous solution of $BaCl_2$[1] to collect them from the knife. Leduc, Marinozzi & Bernhard use the following procedure:

Embedding mixture:

97% GMA/water plus 0.5% (sat.) ammonium persulphate 70 ml

85/15 n-butyl /methyl methacrylate + 1 % benzoyl peroxide 30 ml

Dehydration:

80% GMA in water	15 minutes
97 % GMA in water	15 minutes
50/50 97% GMA/embedding mixture	15 minutes

Place the specimens in capsules containing viscous partly-polymerized embedding mixture and polymerize with strong ultra-violet in cold. The resin will harden overnight. The 30 % methacrylate content of the final embedding medium is used to reduce the very hygroscopic properties of the final GMA block.

Durcupan[2,3]

This is a commercial water-soluble resin which has been used for electron microscope embedding. However, it gives a very soft block and Aquon or glycol methacrylate is far more suitable.

Special embedding mixture

Methacrylate-Styrene

This embedding medium is used in conjunction with the special techniques for preserving damaged, senescent and highly vacuolated plant cells as outlined in the chapter on fixation. The mixture is as follows

7 vol. n-butyl methacrylatel; stabilizer removed.

3 vol. styrenc vinyl benzenel; stabilizer not removed.

Add 1 % (w/v) benzoyl peroxide to the above as a catalyst. First remove the water from the benzoyl peroxide with anhydrous calcium sulphate. The embedding medium should be centrifuged before use.

4

MICROTOMY

Ultrathin sectioning of plant tissue presents problems not encountered with animal material. The most important of these is the vast range of variation in toughness found in a single block of plant tissue. The knife must cut the cell wall which often, as in lignified tissue, is extremely hard. At the same time it must cut the much softer cytoplasm and the resin-filled vacuole. This conflict sets up considerable strain at the cutting face and can easily result in imperfect sectioning of one or more of the components, or in distortion of the section. A further problem can be the presence in the cell of hard inclusions such as starch grains or silicate concretions. These cannot be sectioned and will be roughly fractured, knocked out completely, or pushed through the specimen by the knife edge, tearing the section. Clearly the first of these alternatives is preferable, the second sometimes acceptable, and the third unacceptable.

These problems must be overcome, in part, at the embedding stage. The resin must be uniformly hard throughout, and the hardness must be correct for the tissue concerned. Without this, attempts at sectioning are doomed from the start. Having achieved this, it remains to trim the block to expose the specimen in the required orientation and with a correctly shaped face. Correct technique is also critically important both in preparing the knife and in carrying out the sectioning.

THE KNIFE

Steel knives were once used for ultramicrotomy, but their sharpness was barely adequate and the invention of glass knives in 1950 made them totally obsolete. A later development has been the diamond knife and both types are now in common use. Each has its own advantages and disadvantages. Glass knives are made by the user, are cheap, short-lived and disposable; diamond knives are bought readymade and are expensive, long-lived and (relatively) durable. Which to choose depends on the job to be done and the pros and cons of each are discussed more fully later. However, whatever the project, it is *essential* for the beginner in ultramicrotomy to start with and become experienced in using glass knives before attempting to use a diamond knife. The edge of any ultramicrotome knife is easily damaged both by careless handling and by incorrect use of the microtome. With a glass knife this is of no consequence-it can be thrown away and a new one used. With a diamond knife it would mean regrinding or even replacement, and both of these alternatives are expensive and involve delays of several months.

GLASS KNIVES

The cutting edge of a glass knife is produced simply by fracturing a piece of glass. Not surprisingly, to obtain good knives the conditions of fracturing must be correct. The glass itself must also be free from internal stresses. Float glass, ¼ (6 mm) thick, seems to be the most suitable and easily available, though almost every type of glass known has been used with success, including old tram windows! Most laboratories buy the glass in strips, 1 in. (25 mm) or 1½ in. (37 mm) wide according to the requirements of the microtome. This is by far the most convenient method, although the glass can be bought in sheets and broken into strips.

If you get variable results using float glass try to obtain Belgian Libbey Owens Flat Drawn Plate Glass, 6-7 mm thick. This is the glass used by LKB for their strip supplies.

Almost all electron microscopists now use a machine such as the LKB Knifemaker to break their knives. The LKB Knifemaker first breaks the strips into squares or rhombi, then scores these diagonally and breaks them into triangular knives. The advantages of using such a machine are manifold. All score marks are of exactly the same length and depth. The breaking also takes place under controlled and exactly reproducible conditions. Provided that the machine is correctly adjusted, every knife will be as good as, and possibly better than, the best that can be made by hand.

However, if a knife-making machine is not available it is not difficult to make knives with the simplest tools, although the knives produced will always vary in quality. All that is required are one or two pairs of glasscutters, pliers and a wheel-type scorer. Do not use a diamond scorer, as the score produced is rather irregular, and stresses are set up in the glass when it is broken.

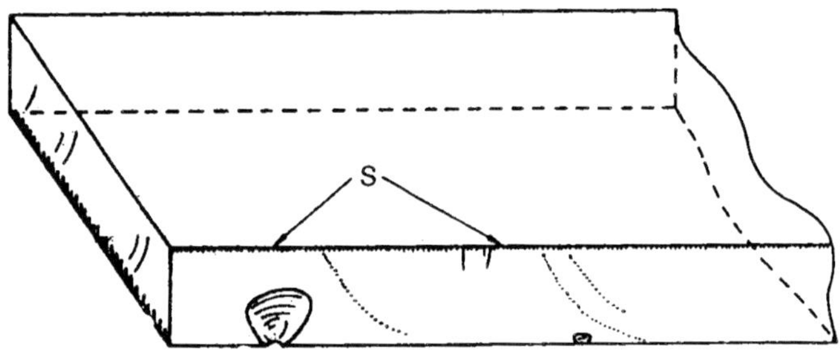

Figure 4.1

Check each glass strip before it is used. The edges are the most important part, since they form the cutting faces of the finished knives. Look especially for stress marks. Some stress marks along one side of the face are inevitable, as they result from the scoring of the glass when the strips were made. These should not extend over more than ¼ of the face; reject any strips in which there are many stress marks larger than this. Discard also severely chipped strips. Strips with a few small chips can be used, provided that the chipped parts are avoided when making (With a knife-making machine, simply discard the squares containing the chips after breaking the strip into squares.) Strips with only one good face can be used; this will mean discarding 50% of the knives if using a knifemaker, but when making knives by hand only one face is normally used. Clean the strips thoroughly with acetone or ether, using tissues or a soft cloth and paying particular attention to the edges.

The position of the score mark required for the first break when making knives by hand; this is the break which produces the cutting edge. Various simple aids to accurate positioning exist. A set-square which locates with the corner of the strip, and has a third arm to give the correct score position, is available commercially. Alternatively, a scoring board can be prepared of clothcovered blockboard, and the correct scoring positions drawn on to it. In use, line up the strips with the marks on the board, and score along a steel ruler placed above the appropriate line.

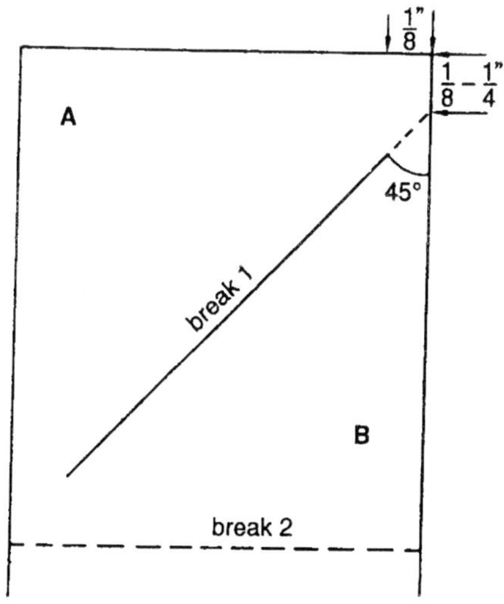

Figure 4.2

It has already been mentioned that there will be stress marks, caused by the original scoring, along the edges of the strips. When scoring the strips to make knives place them with the original score marks (visible as a narrow serrated band along one side of the strip edge) facing downwards; the lower part of the cutting edge is never usable, so it does not matter if it is affected by stresses in the glass. Make the score gently but firmly. If the score is too light the break will not follow it, if it is too heavy, stresses will be set up in the glass.

Three methods of breaking the glass along the score. In A two pairs of pliers are used. The jaws of each pair must be close to and parallel with the score mark. Make the break with a horizontal pull. B uses a modified pair of glasscutters pliers which can be bought as *Knipex* pliers or made by sticking small strips of tape in the middle of one jaw and at the edges of the other. Grasp the strip so that the central piece of tape is on the side away from the score, and accurately parallel to it. If the score has been correctly made, moderate pressure will give a clean break. In C the tip of a 6 mm glass rod is heated to bright red heat in a bunsen and pressed on to the centre of the score. The strip should break cleanly along the score. Remove the knob which forms on the end of the glass rod before using the rod again.

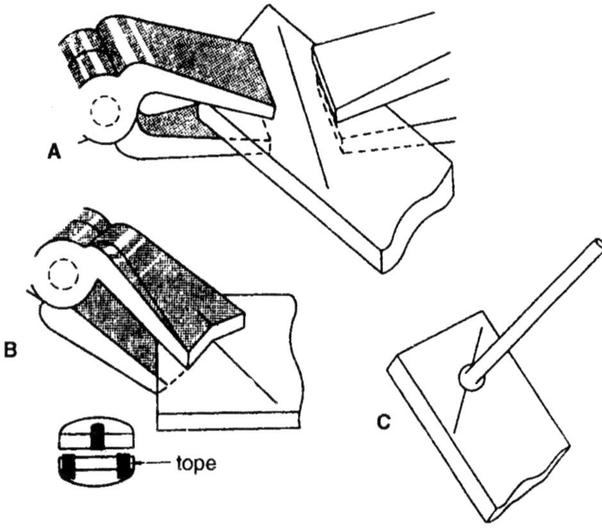

Figure 4.3

To complete the knife make a second score and break. The piece A is the knife. The piece B will not usually be a good knife. However, luck is a major component in knife-making by hand, and it is always worthwhile examining both pieces! Some microtomes require knives which are not triangular-the Porter-Blum[13] requires a parallelogram, the Cambridge-Huxley a trapezium. In these cases the second break must be made in the appropriate position.

Just behind the cutting edge of a knife a stress mark can be seen, if the knife is viewed obliquely. The shape and position of this stress mark indicate the knife's sharpness and useful cutting length. The various possibilities and how they are to be interpreted. The gap between the arrows indicates the useful cutting edge. The edges in Figures 4.4 D, G and H are useless. The straightness of the cutting edge is also a guide to the knife's usefulness. A perfectly level edge is convenient to use, but often not so sharp as a slightly curved one. A very curved one, though, is unlikely to be sharp and would be impossibly awkward to use.

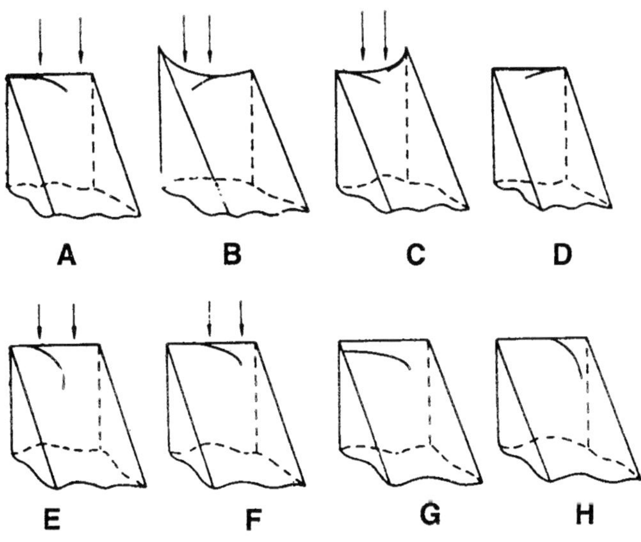

Figure 4.4

For some purposes a knife angle other than 45° is required. More acute angles (30° upwards) deform the section less and may be sharper, but are shorter lived. They are often used for cutting blocks embedded in Polyester resin such as Vestopal, which will not stand much deformation. More obtuse angles, 50-60°, are less sharp but longer lived. They can be useful when cutting serial sections. Whether a change in knife angle would help a particular problem is something that can only be found by experiment but in most cases 45° knives will be found the most useful.

Store glass knives in plastic sandwich boxes, mounted upright with their bases stuck into a strip of plasticine.

Diamond Knives

Since diamond is extremely hard, a knife made from it has a much longer-lived cutting edge than a glass knife. However, diamond is crystalline, not vitreous, and this means that knife-edges cannot be prepared by fracturing, but must be ground. Grinding an edge suitable for ultramicrotomy is a difficult and expensive process, and diamond knives are therefore costly. They are sold at a price per millimetre of usable cutting length, and supplied with cutting lengths of from two to four millimetres. A range of knife angles is available, but as with glass knives, 45° is usually the best compromise for botanical work.

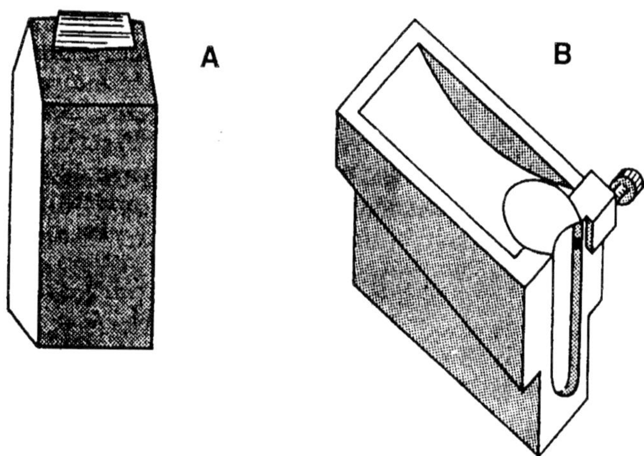

Figure 4.5

The knife itself is a very small chip of diamond and is supplied permanently fixed in a brass strip. A special holder is needed to mount this in the microtome. Use great care when fixing the knife in this holder. Make sure also that the knife is the right way round; the brass strip will be marked to indicate the correct orientation. The holder acts also as a water-bath, so the knife must make a water-tight fit with it. Wax can be used to seal it in, though with a well-made and undamaged holder this should be unnecessary. Do not remove a knife from its holder unless it is essential to do so; keep the knife in its holder in a closed container and place it in the microtome (which it fits in exactly the same way as a glass knife) when required. Never leave the knife in the microtome when not in use.

Although diamond is very hard, it is also brittle and the edge of a diamond knife is very easily damaged. It should never be touched by anything but the block being sectioned, except when it is being cleaned. Correct cleaning is vital to the successful use of a diamond knife. Remove dust with a very fine sable brush, always brushing upwards, towards the edge. Routine cleaning, before each use and whenever the knife becomes dirty during sectioning is carried out with a stick of balsa wood, $\frac{1}{16}$ in. (1.6 mm) square in section. Trim the end of this to a tapering wedge, cutting a fresh end each time the knife is cleaned. Suck this end until it is moist with saliva, but not excessively wet, and run it along the knife edge holding the stick vertically and not using any pressure. Saliva is by far the best cleaning medium-it makes the edge easily wettable and may also leave a protein coat on the knife which lubricates cutting.

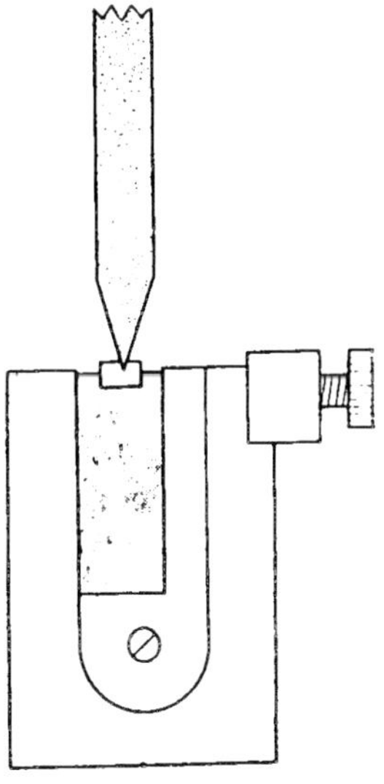

Figure 4.6

Sometimes the knife edge may become covered with an oily coating which the above procedure fails to remove. This calls for a more drastic cleaning method which will only very rarely be needed, although it is particularly likely to be necessary when cutting Araldite. Cut a balsa stick as before and dip the end into concentrated sulphuric acid. Pass it along the edge exactly as in the routine cleaning. Leave it for a few seconds, then rinse the knife under the hot tap, holding it with the cutting edge upwards. Let it stand for 20 seconds, then immerse it in an N/10 solution of sodium carbonate. Rinse it well in cold distilled water, then dry it with a hair dryer or hot-air blower. The edge should face into the air stream, so that water is blown away from the edge rather than evaporated from it.

A diamond knife will eventually need to be reground, and this can only be done by the manufacturers. Knives which see a lot of use and are used to cut tough tissue may well need regrinding after only a few months of use. It is unlikely, therefore, that a diamond knife will be cheaper to use than glass knives. A diamond knife is invaluable, though, for cutting wood and similar very tough tissues. It is also worthwhile if a lot of serial-section work is done, since sectioning can continue indefinitely without the need to change the knife position periodically. With some thickened tissues a diamond knife is essential for serial sectioning, since only one or two sections can be obtained from any one part of a glass edge.

WATER BATHS

Electron microscope sections are very delicate, since to be usable they must be between 50 and 150 nm thick. Such sections cannot be cut dry-there must be a water bath behind the knife, with a water surface extending right to the cutting edge, so that each section can float away as it is cut. A 'boat' is therefore fixed to the knife to contain the water.

Figure 4.7 shows two types of boat used with glass knives. 9A is made of thin sheet metal and fixed and sealed to the glass with paraffin wax, applied with a hot spatula. 9B is simply made from a strip of PVC tape, bent in a loop round the back of the knife. The bottom of the loop is sealed with paraffin wax. Make sure in each case that the boat reaches the top of the knife, or it will be difficult to wet the cutting edge. Diamond knife holders always incorporate a water bath. The water bath of some holders is inconveniently large when cutting very small sections and a hydrophobic barrier can be used instead. This consists of a strip of Durofix adhesive, or a line drawn with a Chinagraph wax pencil, placed across the holder ¼ in. (6 mm) behind the knife edge. A drop of water, on to which the sections are cut, is retained between this barrier and the knife edge.

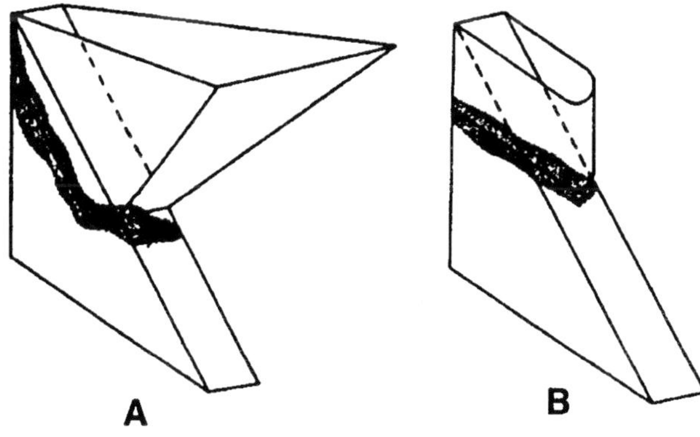

Figure 4.7

The water bath does not always contain pure water. Methacrylate sections must be cut on to a 10-15 % solution of alcohol or acetone. This softens the sections, and enables any wrinkles resulting from the strain of sectioning to flatten out. Epoxy resins are not affected by solvents and can be cut on to pure water. However, a 10% solution of alcohol or acetone is frequently used since it has a lower surface tension and therefore wets the knife edge more easily. Water-soluble embedding media present a different problem. In fact most are only very slightly soluble after polymerization and can be cut on to water, which has the same softening and flattening effects as an acetone solution on methacrylate. To avoid excessive softening some workers use 30-40% acetone when cutting Durcupan. This, however, has a very low surface tension and the sections tend to dart about, making them difficult to pick up.

When filling the boat, with the knife *in situ* on the microtome, first add an excess of water to give a convex meniscus. This makes sure that the knife edge is wetted. Then gently withdraw water with a hypodermic syringe or Pasteur pipette, while looking through the binocular viewing microscope. The illumination system is so arranged that, as the meniscus becomes concave, a bright reflection from the light source will be seen in it. This is the required water level and the bright reflection will make it easy to see the sections as they come off. With some diamond knives it will not be possible to make a sufficiently convex meniscus initially to wet the knife all along its edge. Instead, sweep the water up to the edge with a very fine sable brush, brushing always upwards. Drops of water will have to be added from time to time during sectioning to maintain the correct meniscus, especially when using small boats. Never allow the edge of a glass knife to dry out once it has been wetted-it will become unwettable and have to be thrown away.

TRIMMING OF BLOCKS

Blocks are trimmed for two reasons : to expose the specimen, and to present a suitable face to the knife. Normally the specimen is only a short distance beneath the surface of the resin if the block has been made in a gelatine capsule. The best procedure then is

1. File away the block around the specimen, leaving a square nub 2-3 mm high containing the specimen. This should be positioned so that the required part of the specimen is displayed in the desired orientation. This stage can be eliminated if specially-shaped 'pretrimmed' polythene capsules are used.

2. With the block in its holder from the microtome, trim the nub to the shape of a pyramid and remove the surplus resin from above the specimen using a grease-free razor blade. This must be done under a binocular dissecting microscope. The tip of the block will now have the form of a truncated pyramid, the top face of which should be square or trapezoidal. The presence of two parallel sides, set parallel to the knife edge, ensures that when the sections are cut they form a straight ribbon on the water surface.

The face should be as small as possible-not more than 1 mm square unless it is absolutely unavoidable. When cutting difficult tissues, and blocks embedded in polyester resins, the face should be less than 0.2 mm square, especially when using a diamond knife. In particular, when cutting non-lignified tissues with very thick cellulose walls (such as collenchyma or stipe cells of brown algae) with a diamond knife, the face must be no more than 0.1 mm square at the very outside.

Sometimes the tissue sinks through the resin during embedding into such a position that to trim it to the required orientation would mean trimming at an extreme angle. This is undesirable, since the microtome advance will then cause lateral movement of the specimen relative to the knife. It may also aggravate vibration problems. When a particular orientation of the specimen is essential it is often advisable to embed the tissue in larger resin blocks, such as the compartments of a polythene ice-cube tray. Pieces can then be cut out to give whatever specimen orientation is needed.

If this has not been done, rather than trim a block at an absurd angle, cut out a small block containing the specimen, and re-embed this. Epoxy resins are not softened by unpolymerized resin, so the same resin can be used for re-embedding. Methacrylate would be softened and the block ruined by this procedure, so the excised block should instead be cemented to a short piece of Perspex rod with ester wax.

SECTIONING

Insert the specimen holder into the microtome so that the parallel sides of the face are parallel to the knife edge. If the face is rectangular, have the longer sides parallel to the knife. Advance the knife gently towards the specimen. First use the coarse advance, watching from the side with the naked eye until the knife is as close to the specimen as is comfortable. Then use the fine advance, viewing through the binocular microscope. On some microtomes it is possible to see a reflection of the knife edge in the block face. In this case, advance the knife until the dark gap between the knife and its reflection disappears. The knife is then very close to the block. Alternatively, the microtome may have a binocular system which can be set to give a vertical view of the gap between the knife and the specimen. Advance the knife until no gap is visible. If a microtome does not have a built-in binocular it is worthwhile when providing the viewing system to ensure that it can be set to give this vertical view.

The last 5 µm or so of advance will have to be carried out 'blind'. If the microtome has neither of the aids mentioned, a greater part of the movement will be blind. Move the specimen round its cutting path, advancing ½ –1 µm each time, until cutting begins. Then leave advancing to the automatic thermal or mechanical mechanism.

The sections that have been cut will be visible as they float on the illuminated meniscus. If all is well they will stay together, forming a ribbon of sections running back from the knife edge. On a manual microtome some practice is needed to maintain the steady rhythm required to achieve this, but there should be no problem with a motor microtome if the block has been correctly trimmed.

Make the following adjustments to obtain satisfactory sections

1. Give the knife a slight forward cant. Its angle from the vertical-the clearance angle-should be about 6° for a glass knife and about 1° for a diamond.

2. Adjust the cutting cycle, or the height of the specimen or knife so that cutting occurs at the correct point in the cycle. This should be just over half-way through the controlled speed cutting movement.

3. Set the cutting speed at 3-4 mm/sec for a glass knife or 1-2 mm/sec for a diamond.

4. Section thickness will have a separate control if the microtome has a mechanical advance, or will be set by a combination of temperature and speed controls if the advance is thermal. In the latter case, changing the cutting speed will affect the section thickness and either the return speed (if there is a two-

speed drive) or the temperature will have to be adjusted to compensate. Sometimes this compensation is automatic.

With either type of advance the manufacturers' indicated section thickness is only a rough guide. Useful information about section thickness can only be obtained from the interference colours seen as the sections float on the illuminated meniscus. Approximate values for the thickness of sections of various colours are given in Table 4.2 below. The colour change is progressive rather than abrupt, so that, for example, a silver-grey section will be around 60-70 nm, a silver-gold section 90-100 nm, and a reddish-gold section around 150 nm.

TABLE 4.2. INTERFERENCE COLOUR AND SECTION THICKNESS

COLOUR	THICKNESS (nm)
Grey	<60
Silver	60–90
Gold	90–150
Purple	150–190
Blue	190–240
Green	240–280
Yellow	280–320

Sections for the electron microscope must be thinner than 150 nm, so only grey, silver or gold sections will be suitable. For high-resolution work sections of at least silver quality are needed.

FAULTS IN SECTIONS

Compression

Methacrylate sections are liable to compress under the strain of cutting. Sections are shortened in the direction of cutting so that, for example, circular structures appear elliptical. In extreme cases the sections may be slightly wrinkled. The acetone solution in the water bath usually softens the sections enough for them to flatten and lose their compression. If this is ineffective, soak a small piece of cotton wool in xylenel and hold it above the sections long enough for them to expand; often a sudden expansion can be seen when watching through the binocular microscope. Epoxy and polyester resins are less liable to this fault and, when it occurs, it often indicates that the resin mixture is too soft. Try chloroform vapour to flatten epoxy sections.

Distortion and Wrinkling

Effects like those caused by compression are sometimes seen, for opposite reasons, when cutting very rigid blocks, especially blocks embedded in Vestopal. The cause seems to be deformation of the section by the bending forces involved as it is cut. One cause can be a dirty knife; the dirt acts as a brake on the section as it slides away from the edge. When this is not the case reduce the clearance angle and use knives with a smaller knife angle. This problem is much more common with diamond knives than with glass and is the reason for using a clearance angle of only 1°. The greater rigidity of the diamond edge is probably the cause; the edge of a glass knife is very flexible and can 'give' to relieve stresses. The only cures available with a diamond knife, if the clearance angle has already been reduced to the minimum, are to reduce the area of the cut face and the cutting speed.

Chatter

Chatter is a variation in section thickness occurring as regular bands parallel to the knife edge. It happens under conditions similar to those causing distortion and is particularly common when using very rigid blocks or diamond knives. The cause again seems to be that the plasticity of the section or the knife is insufficient to allow stresses to be relieved. A combination of chatter and deformation can cause rippling-parallel folds extending across the whole width of the section. If chatter occurs when sectioning normal blocks with a glass knife, suspect first that vibration is affecting the microtome. The microtome must stand securely on its table, which must in turn be on a solid floor. Do not touch the microtome or table while sectioning is in progress, except to operate the lever or wheel if it is a hand microtome. Vibration may also be set up inside the microtome by the motor or the pivot bearings, and if this is suspected the machine will need expert attention.

When vibration is not the cause, the cure will usually be found by reducing the cutting speed. However, the cutting stroke of most microtomes becomes jerky if the speed is reduced too far and this sets a limit to the usefulness of this cure. Reducing the knife and clearance angles, if this is possible, can also help. Very thin (grey) sections are often easier to cut without chatter than thicker ones, when using a diamond knife.

Knife Marks

These are scratches or tears running in the direction of cutting and are caused by imperfections in the knife edge. The remedy is to use a different part of the edge. Minor knife marks, causing merely a difference in section thickness in a strip along the direction of cutting, cannot always be avoided and may be useful in indicating the direction of cutting when the section is in the electron microscope.

Sections of Varying Thickness

1. When alternate thick and thin sections are cut the cause is usually a poor knife. However, a thick section cut by accidental over-advance of the knife will usually be followed by a very thin one, and this can set up a thick-thin sequence when cutting is first started. This should settle down progressively to normal cutting.

2. If the sections do not alternate, but vary randomly in thickness, the cause is normally temperature variation (often caused by draughts), or, with a hand microtome, failure to keep a steady rhythm.

3. When no sections are cut for several cycles, and then a very thick section is suddenly cut, the cause is a negative clearance angle. The block is being progressively compressed for several cycles and then finally springing out far enough to be cut.

4. A sudden decrease in thickness indicates that the microtome has run out of advance and must be reset.

Sections of Uneven Thickness

These may be caused by a knife which is not uniformly sharp, or by a block which is irregularly polymerized. A 'soft-centred' block can often be improved by heating in an oven at 60°C for 48 hours. Never re-heat a block with its gelatine capsule still in place. The first few sections cut from a block which has been trimmed with a razor-blade will always be uneven, since a razor-blade produces a very coarse cut by the standards of ultramicrotomy.

Failure to Cut

If no sections are cut at all, the cause is probably one of the following faults:

1. Knife completely blunt or unsuitable part of the edge in use.
2. Block not penetrated by resin or resin very soft.
3. Knife or block not securely fixed in microtome.
4. Pronounced negative clearance angle.
5. Block face wet. This may simply be due to spilt water, in which case dry it and the face of the knife with fluffless filter paper. It may, on the other hand, indicate that the resin has not penetrated properly and the hollow cells are taking up water by capillary action. In this case the block is useless unless it can be re-trimmed to expose a properly embedded portion. An overfull boat can also cause wetting of the block face, but this should not happen if the meniscus has been correctly adjusted to give reflection of the light source.

MOUNTING

Sections are mounted on metal grids for viewing in the electron microscope. These are normally made of copper, but other metals are used if copper could interfere with staining reactions. Depending on the microscope they are intended for, the grids will be 2.3 or 3 mm in diameter. The grids used may either be of coarse mesh (50-100 meshes/inch, 20-40 meshes/cm), coated with a thin carbon or plastic film to support the section, or fine mesh (200-1000 meshes/inch, 80-400 meshes/cm), used without a film. Methacrylate sections always need a support film, since they have a lower mechanical strength than epoxy or polyester sections and also tend to volatilize in the electron beam.

If the sections can be mounted without a film, the choice lies between higher resolution and better contrast (fine mesh without a film) or less of the section obscured by grid bars (coarse mesh with a film). The use of coated grids demands higher standards of cleanliness, since the support film provides an additional source of contamination by dirt particles. Clearly, the choice will depend on the project in hand. The preparation of the various types of support film is described.

Only pick up good sections on the grid. When cutting, wash bad sections over the edge with a squirt of water or sweep them over with a sable brush. The microtome has a receptacle beneath the knife for water swept over in this way; this must be dried out after use.

A tool for moving sections on the water surface can be made by taping an eyelash to a thin metal or glass rod. Use an eyelash in preference to a hair since it has a pronounced taper from root to tip, giving it both rigidity and a fine tip. With this tool it is possible, with practice, to move sections about freely on the water surface, so that if a ribbon disintegrates, the parts can be moved together again and picked up on one grid. Move the sections away from the knife edge before picking them up; there is little point in this if the edge of a glass knife has already become blunt, but there is every point when using a diamond knife, since damage to the edge can be so costly. Carry out all these operations and the picking up of the sections, under the binocular microscope.

Take a grid by the edge in a pair of fine forceps and press it gently against a piece of filter paper so as to give the matt side (or the coated side) a slight convex curvature. Carefully touch this side of the grid on to the sections in the water bath. They will adhere to it when it is lifted off. Place the grid, matt side up, on a piece of flufliess filter paper in a petri dish. The grid can be left in the dish, or, if it is to be transported, placed in a grid storage box.

If grids are handled a lot after the sections have been picked up, construct the simple manipulator shown in Figure 4.8 to avoid the risk of damage by forceps. A source of low vacuum such as a tap pump or small cylinder pump, is connected to one broad branch, the other is covered by the thumb to control the vacuum and the fine branch is used to pick up the grids; a ready made device is available commercially. Do not use it, however, for picking up sections from the bath, nor for putting grids through staining solutions.

A sequence of serial sections of an object can often give valuable information about its three-dimensional structure. Grid bars across the sections can be very annoying when examining serial sections. If it is not essential to see every section use 50 mesh grids with supporting films, but often it is preferable to use, rather than grids, discs with one or more large holes. Coat these with a double Formvarcarbon film. They can also be useful when examining large single sections by mosaic photography.

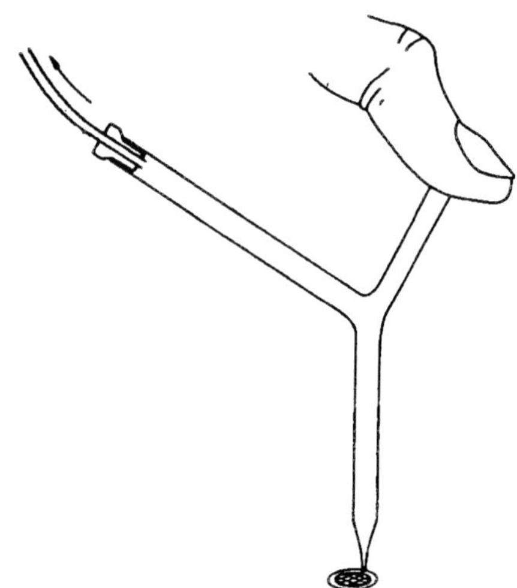

Figure 4.8

Grids with slots instead of holes are sometimes useful, but the ribbon of sections must be very straight and a special mounting technique is needed. Pick up the sections from the trough, not on a grid, but on a loop of thin wire, 4 mm in diameter, covered with a Formvar film. (To prepare this, dip the loop in a dilute Formvar solution in chloroform, drain vertically, and allow it to set.) Place a 'slot grid' on a Perspex rod' and lower the loop on to it, lining up the ribbon of sections with the slot, until the film is transferred to the grid. It needs a very steady hand to do this freehand and a micromanipulator is recommended. Barnes & Chamber (1961) describe a very simple one which can easily be improvised in the laboratory.

Various special grid patterns are available for particular purposes. Hexagonal grids give the maximum support for a given mesh size. Diamond patterns are useful if a knowledge of north-south orientation is essential. Rectangular patterns also provide this information, and can reduce interference from grid bars if long structures running in a particular direction are being studied. If it is often necessary to find a particular part of a grid, grids with numbers in the squares are obtainable. These are particularly useful when grids are to be used for demonstration purposes.

5

STAINING

CHEMICAL COMPOSITION OF DYES

It is best to give the general name of 'colouring agents' or 'colorants' to substances used in microtechnique to colour or blacken the parts of organisms.

Colouring agents are used in very diverse ways. It is a strange fact that the housewife is more careful in her terminology of colouring agents than many microscopists are. She distinguishes clearly between *staining* and *painting* the floor, while they often use the word 'staining' without regard for the diversity of the processes grouped by them under this single name. The 'staining' of specimens by electron-microscopists does not involve dyeing. The word was formerly used as a synonym for 'dyeing', but has come to be treated so loosely in microtechnique that it is avoided in this book.

There are five principal ways of using colouring agents to distinguish the microscopical parts of organisms. In this book we are primarily concerned with only one of these, namely, *dyeing;* but this one is best understood by contrast with the others. The five will be briefly described here.

Injection of suspended coloured particles into closed spaces. We may suspend minute, insoluble, coloured particles in a fluid and inject this into the blood-vessels or other internal spaces of an organism. The final distribution of the coloured particles will be determined by their incapacity to penetrate the walls of the vessels into which the fluid has been forced. The chemical nature of the colouring agent is irrelevant, provided that it is insoluble.

Uptake of suspended coloured particles by phagocytic cells. We may provide a phagocytic cell with minute coloured particles suspended in the fluid in which it lives, and allow it to eat them. The distribution of the coloured particles at any time will be determined by the vital activity of the cell (movements of the cytoplasm, & c.). The particles will not colour any pre-existent object in the cell.

Solution of a lysochrome. We may dissolve a lipid-soluble colouring agent (not a dye) in 70 % ethanol or some other suitable medium, and soak a section of fixed tissue in it. The lipids of the tissue will take up the colouring agent simply because it is more soluble in them than in 70% ethanol, and the distribution of the colour will be determined by this. Such colouring agents are called lysochromes, because they colour by solution (Greek *lúsis,* solution).

Local formation of a coloured substance. We may soak the fixed tissues in a solution of a substance that will react with one or more of the tissue-constituents to produce an indiffusible coloured product; the solution used will be colourless or differently coloured from the product. Thus a yellow solution of potassium ferrocyanide will give a blue precipitate of insoluble Prussian blue wherever there is ferric iron in the tissue. The distribution of the colour will depend on the chemical properties of the reactive substance or substances in the tissue. Many histochemical tests fall into this category of colouring reactions.

Dyeing. We may soak tissues (living or dead) in a solution of a dye. Certain tissue-constituents will combine with the coloured ions of the dye and thus become coloured. The colour will usually not change. The final distribution of the dye will depend on the ability of the dye to penetrate the tissues and on the affinity of the tissue-constituents for its coloured ions.

The nature of dyeing is only briefly indicated by what has just been said, but it will be clear that the process is entirely different from the first three described above. It shows some resemblance to the fourth, but there are important differences. Dyes do not usually change colour when their ions are taken up by the tissues;

and the uptake is generally far less specific, for most parts of the cell are capable of being dyed to some extent by most dyes, if *sufficient* time is allowed.

For the purposes of microtechnique, dyes may be defined thus. They are aromatic, salt-like, crystalline solids, that dissolve in water or aqueous solutions in the form of ions; either the cations or the anions (occasionally both) are coloured; the coloured ions can link themselves chemically with proteins (and generally with other tissue-constituents as well); when the linkage takes place, the ions do not lose colour, and usually they do not change it.

Two questions present themselves. What causes dyes to be coloured, and what causes their coloured ions to attach themselves to tissue-constituents?

Many of the familiar colours of nature—the brilliant wings of many butterflies, for instance, and the iridescent feathers of birds—owe their colour to parallel plates or parallel striations, separated by distances varying from about a quarter of a wave-length to a few wave-lengths of visible light. These 'structural' colours disappear when the substance is dissolved, because they are not due to any colour intrinsic in the molecules or lesser particles of the substance. It is only when such particles themselves show a differential transmission of light of different wavelengths, that a substance remains coloured on solution.

If all substances that remain coloured on solution are examined, it is found that most of them fall into a few major groups. It is a familiar fact that the salts of chromium, iron, and cobalt are coloured, and that colour is retained when they are dissolved. These are only examples of the wide generalization that the ions and ionic complexes formed by the transition elements are coloured; or, to dig a little more deeply into causes, we may say that the elements that give coloured ions are those metals that possess an incomplete shell of electrons inside their outermost shell. That is to *say,* the cause lies in intra-atomic structure. No dye, however, owes its colour to the possession of a transition element, nor indeed to the possession of any particular atom as such: its colour is due to its inter-atomic, not its intra-atomic structure.

Many aromatic, but few aliphatic compounds are coloured; and this suggests that there must be something in ring-structure that favours the production of colour: that is to say, that favours the absorption of light of particular wave-lengths. This is true; and all dyes are aromatic compounds. Benzene itself, the simplest of this group of substances, would appear coloured if we could see a short way into the ultra-violet, for it has an absorption-band at the wave-length of 256 mμ. The molecule may be thought

Alternative structural formulae for benzene

of as 'resonating' between different molecular configurations; that is, undergoing rapid change from one to another.

The alternation of single and double bonds between carbon atoms is favourable to resonance and the associated absorption of electro-magnetic waves; but there are certain molecular arrangements that excite the molecule to a particular degree, so

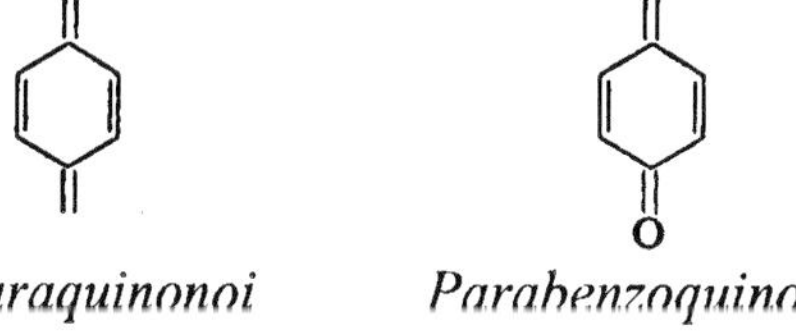

The paraquinonoi ring *Parabenzoquinone*

that light of greater wave-length is affected, and colour results. A very frequent arrangement of this sort in dyes is the quinonoid. All substances possessing a paraquinonoid ring are coloured.

Parabenzoquinone is the simplest of such substances. It is a yellow solid, dissolving in water to give a pale yellow solution, but it is not a dye. It dissolves to form a solution of molecules, whereas dyes dissolve as ions. Parabenzoquinone has no special tendency to attach itself to proteins or other tissue-constituents in such a way as to impart colour to them: dyes have such a tendency, powerfully developed.

The structure of dyes may be illustrated by the one known commercially as *pararosaniline.* (The *para* in this word is not used in its chemical sense.) This is a magenta solid, easily soluble in water. To make it, one needs aniline and paratoluidine. Aniline

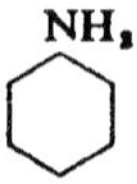

NH_2 NH_2

Aniline 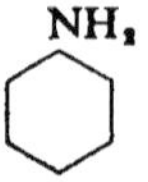 *Aniline*

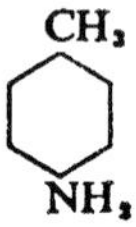

CH_3

NH_2

Paratoluidine

NH_2 NH_2

C

NH_2^+ Cl^-

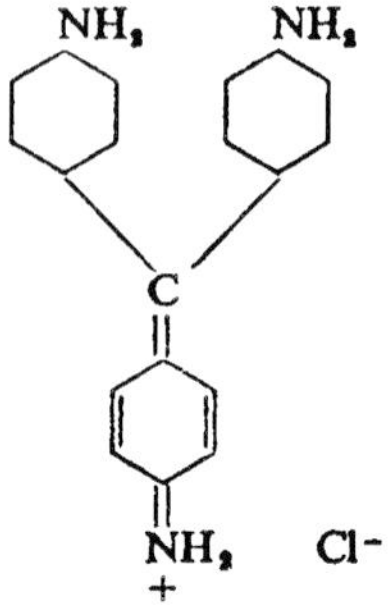

Pararosaniline

is a colourless fluid (if pure), paratoluidine a colourless, crystalline solid. Heat them in the presence of a mild oxidizing agent and a source of chloride ions, and a brilliant colour develops. Two molecules of aniline combine with one of paratoluidine to give a substance in which one of the three rings is quinonoid.

To which of the three rings is colour due: to those derived from aniline, or to that derived from paratoluidine? The formula

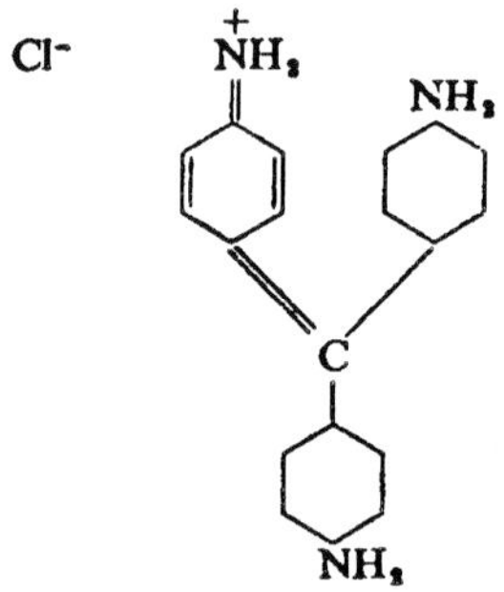

Cl^- NH_2^+

NH_2

C

NH_2

Pararosaniline: another resonance position

given first suggests the latter, but other formulae show the structure equally well. In fact, there is resonance between different structures: one may think of the positive electric charge as being now in one position, now in another.

Resonance is favoured, and the intensity of colour thus increased, if there is a regular alternation of single and double bonds in the coloured ion of a dye. This alternation is called 'conjugation'. Curved arrows are used in the structural formulae of dyes to call attention to conjugation. The structural formula of Biebrich scarlet is shown in figure, as an example of a highly conjugated dye.

Pararosaniline possesses a quinonoid ring, which gives it colour, and an electrically charged group, which enables it to attach itself to proteins and other tissue-constituents. The quinonoid ring is the *chromophore* or colour-bearer; the -NH_2 groups, all capable of conversion to =NH_2, are the auxochromes or colour-helpers, which help the chromophore to attach itself, and often greatly increase the intensity of the colours. The substance as a whole has the general character of a salt.

There are not very many different auxochromes. Many dyes have the same auxochrome as pararosaniline, namely $=\overset{+}{N}H_2$.Since this group of atoms is positively charged, the coloured ion is the cation and the dye is called *cationic* or *basic. The* anion may be chloride or sulphate or acetate; any inorganic anion will do, provided that it gives sufficient solubility. Such dyes may be represented by the formula R^+Cl^-, if the anion is chloride. The hydrogens of the $=\overset{+}{N}H_2$ group are often substituted by methyl or ethyl, or by aryl (the latter being a benzene ring with one or more substitutions of hydrogen atoms).

In many dyes it is the negatively charged ion that is coloured; these dyes with coloured anions are called *anionic* or *acid.* Their cation is usually sodium or potassium. If it is sodium we may write the formula Na^+R^-; R^- means the coloured anion. Various auxochromes give the necessary negative charge. The most usual are the sulphonic, hydroxyl, and carboxyl groups, which ionize to give $-SO_3^-$, $-O^-$, and $-COO^-$ respectively.

It is to be noticed that the so-called 'acid' dyes are seldom acids. The possession of the potentially acidic groups (sulphonic, hydroxyl, and carboxyl) gave rise to the name. Such groups are called acid radicles, despite the fact that they can only form part of an acid if the cation is hydrogen. In a few dyes this is so; the simplest formula for such dyes is H^+R^-.

Beyond auxochromes and chromophores, dyes often possess atoms or groups of atoms that are called *modifiers*. Thus rosani

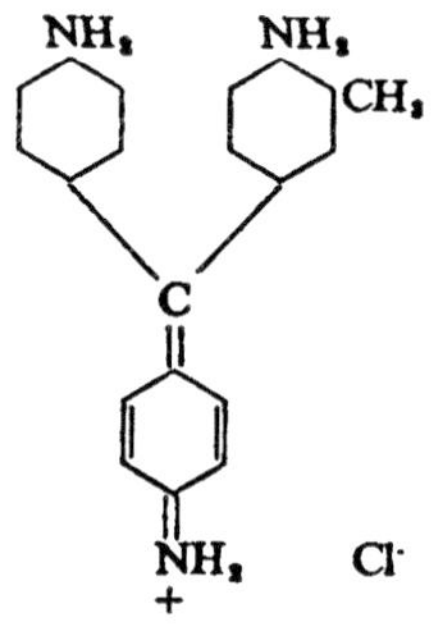

Rosaniline

line differs from pararosaniline only in the possession of a single methyl group; this modifies the colour, making it very slightly bluer. (Basic fuchsine is a mixture of these two closely related dyes.) The methyl group is here a modifier. It has been mentioned that the hydrogens of basic auxochromes may also be replaced by methyl or ethyl or aryl groups; this again modifies the colour. Each replacement of one of these hydrogens in pararosaniline by methyl makes a bluer dye, and crystal violet, in which all six are replaced, is nearly blue. Ethyl and particularly aryl groups have even more blueing effect, and purely blue dyes such as methyl blue may be obtained in this way.

Dyes are classified into groups by their chromophores. Most of these groups contain both cationic and anionic dyes, possessing the auxochromes mentioned above; many of them in all groups have modifiers (frequently methyl or ethyl). The great majority of the dyes used in microtechnique owe their colour either to the quinonoid ring or to the azo-group, –N=N–. The quinonoid dyes, however, are so diverse that it is necessary to divide them into sub-groups, each characterized by a particular chromophore that can be represented in a skeleton formula. The principles of dyeing in microtechnique can be explained by the use of only a few dyes, belonging to a small number of groups. We shall concern ourselves with only five sub-groups of quinonoid dyes. All the dyes mentioned in this book are listed here for convenient reference, under the names of their groups and sub-groups.

Dyes are not known by their full chemical names, because these would be inconveniently long. Shorter names are used, many of which are to some extent descriptive. Initial letters, written in capitals, often form part of a name; these serve to distinguish a dye from others that are closely related. Thus SS means spirit-soluble (soluble in ethanol, not in water); G means *gelb* (yellow). The B of azure B is an arbitrary mark of distinction from two closely related dyes called azures A and C.

QUINONOID DYES

Triarylmethane

Cationic. Pararosaniline, rosaniline, methyl violet, crystal violet, aniline blue SS

Anionic. Methyl blue

Haematein

Anionic. Haematein

Anthraquinonoid

Anionic. Alizarine, purpurine, carminic acid

Xanthene

Anionic. Eosin

Thiazine

Cationic. Thionine, azure B

AZO DYES

Mono-azo

Anionic. Orange G
Dis-azo
Anionic. Biebrich scarlet

QUINONOID DYES

Triarylmethane dyes. These may be regarded as derived from leuco-pararosaniline, which is a colourless substance, since it possesses no chromophore. Leuco-pararosaniline itself is methane in which three of the four hydrogens have been replaced by aniline. The aniline rings are merely particular examples of aryl

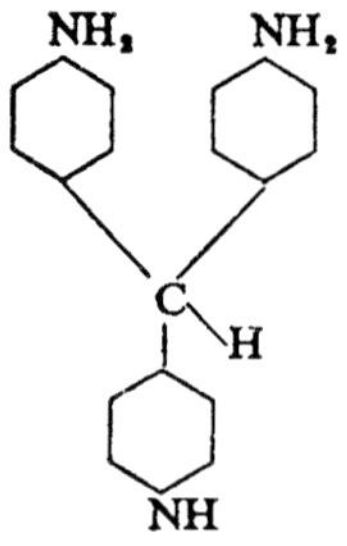

Leuco para-rosaniline	*Skeleton formula for triarylmethane dyes*	*Triphenyl para rosaniljne*

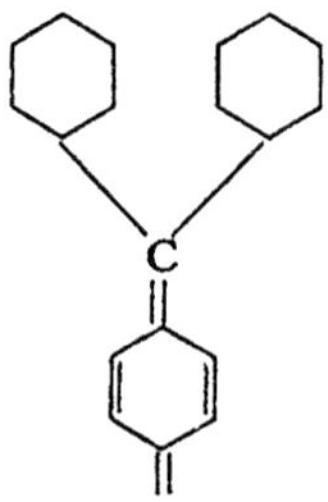

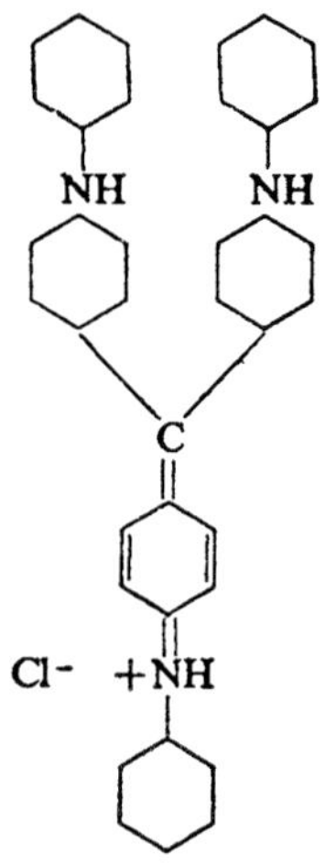

rings, and all the many dyes that have three such rings held together in the same way, whether any of the rings has an $=\overset{+}{N}H_2$ group on it or not, are therefore classified together as triarylmethane dyes. In the skeleton-formula for this and other groups of dyes, the auxochromes and modifiers are omitted, since these differ from dye to dye.

Methyl violet is pararosaniline in which four or five of the hydrogens of the amino-groups have been replaced by methyl. This replacement has a strong blueing effect. When all six are replaced, crystal violet is produced; this dye is almost blue. It has already been mentioned that ethyl and aryl groups have still greater blueing effect than methyl. This is well exemplified by triphenyl-pararosaniline, which, with another closely related dye, constitutes the mixture known as aniline blue SS. Triphenyl-pararosaniline is a pure blue cationic dye. It is pararosaniline in which one of the hydrogens of each of the three amino-groups has been replaced by a phenyl group (that is, by a simple, unsubstituted aryl group).

Methyl blue is a pure blue anionic triarylmethane dye that owes its colour to three extra aryl groups acting as modifiers. Despite its misleading name, it possesses no methyl group.

Haematein. This is a very small group of dyes that are derived from natural products and are not made synthetically. As the

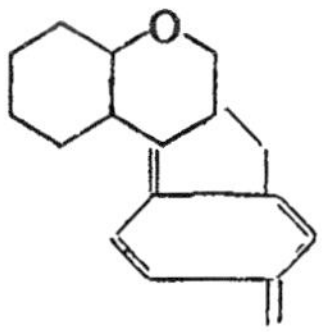

Skeleton formula for haematein dyes 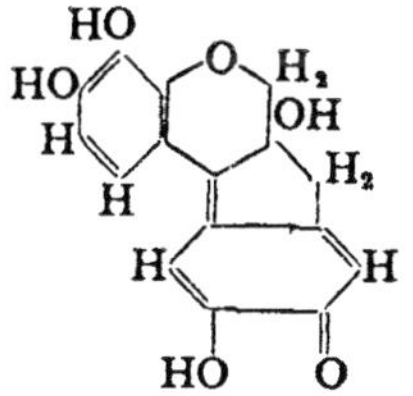 *Haematein* 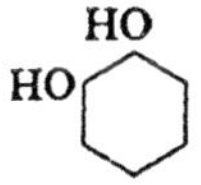 *Catecho*

formula shows, this is not only an acid dye, but actually an acid, related to catechol. It is of a brownish-red colour, but becomes blue in alkaline solution. It is therefore used as an indicator of pH, but it is not valuable by itself as a dye. When used with an intermediary or 'mordant' between itself and the tissues, it becomes one of the most important dyes used in microtechnique.

Haematein is derived from the heart-wood of a small leguminous tree, *Haematoxylon campechianum,* native to Central America. This contains a non-quinonoid substance, haematoxylin, which is not a dye and is indeed colourless, since it lacks a chromophore; but it is readily oxidized to haematein by weak oxidizing agents, including atmospheric oxygen. Haematoxylin is often sold in a partially oxidized form, as a brown powder that is really a mixture of haematoxylin and haematein.

Anthraquinonoid dyes. These are related to anthraquinone, a yellow, crystalline substance. The simplest is alizarine, the chief coloured constituent of madder, a vegetable dye. Alizarine is easily synthesized from anthraquinone. It is seldom used in microtechnique. Its complex relative, carminic acid, is a particularly

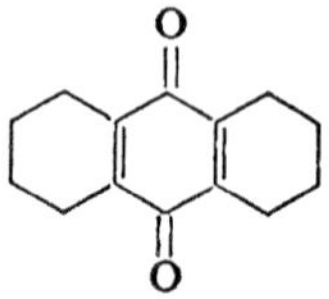

Anthraquinone 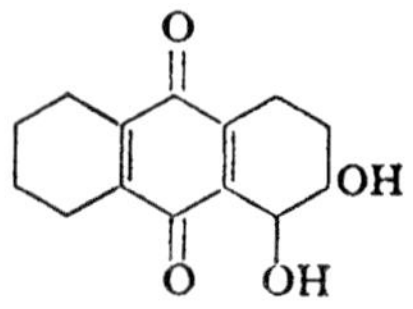 *Alizarin*

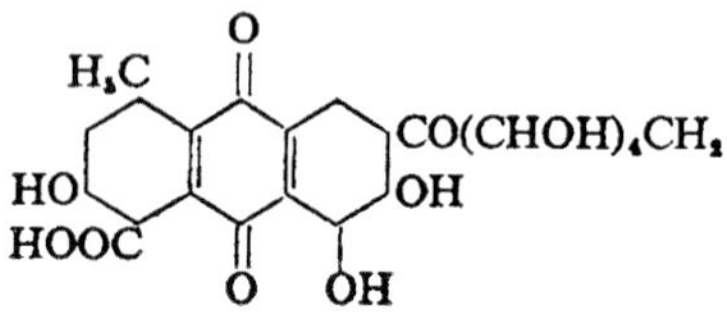

Carminic acid

valuable dye in biological work. Carminic acid is soluble (unlike alizarine) in distilled water, and also in ethanol. It occurs naturally in the fat-body of the wingless females of the coccid *Dactylopius cacti*. This is a scale-insect that sucks the juices of the succulent plant *Nopalea coccinellifera*. The latter is cultivated in various subtropical parts of the world to provide the insect with food. The dried females constitute cochineal. Carmine is a crude form of the dye, produced by precipitating an aqueous extract of cochineal with alum (potassium aluminium sulphate). It is composed mainly of carminic acid bound to aluminium and to protein derived from the insect. Carmine is insoluble in distilled water; but solutions can be obtained (e.g. by the addition of acid or alkali), and the crude dye is usable in microtechnique. It is best to use the pure acid whenever one wants to know exactly what reactions are occurring.

One of the main uses of carmine in microtechnique is explained in the chapter on the use of mordants.

Xanthene dyes. In these dyes, carbon and oxygen form links between two rings. Not many members of this group are commonly used in microtechnique. Eosin, however, is particularly familiar. It is a valuable 'background' dye: that is to say, it is anionic, pale and diffuse in action; it is used to give contrast with objects picked out vividly in another colour by another dye.

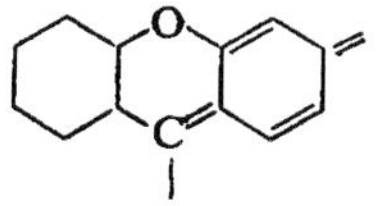

Skeleton formula for xanthene dyes

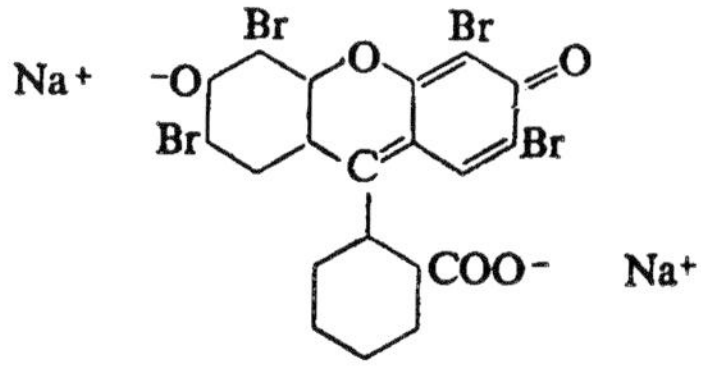

Eosin Y

The name refers to the pinkish-orange colour of the sky at dawn (Greek *eos,* dawn). It will be noticed that eosin has two different auxochromes, and that bromine is here used as a modifier. In related dyes chlorine and iodine partially or wholly replace the bromine, and a bluish tinge is thus imparted.

Thiazine dyes. Here sulphur and nitrogen link two rings (Greek *theion,* sulphur; French *azote,* nitrogen; the *a* in thiazine is short, as in *azote).* Thionine, a blue dye with a reddish tinge, is the

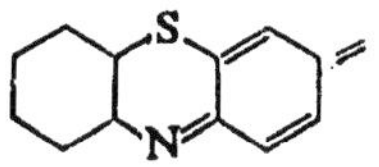

Skeleton formula for thiazine dyes

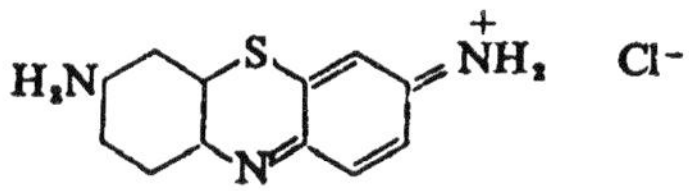

Thionine

simplest member of the group. The thiazines used in micro-technique are all cationic. They are valuable dyes for nucleo-proteins. Several of them, including thionine, are 'metachromatic' ; that is to say, they impart quite different colours to different substances. This is a most valuable property, since it gives important indications of chemical composition. Azure B is another metachromatic thiazine dye. It is trimethyl thionine; that is to say, thionine modified by the substitution of three methyls for three of the four hydrogens of the aminogroups.

AZO DYES

These contain no quinonoid ring, apart from a few that have both azo and quinonoid chromophores. The name refers to the ni-trogen atoms that link two rings together. (The name is derived from French *azote;* short *a.)*

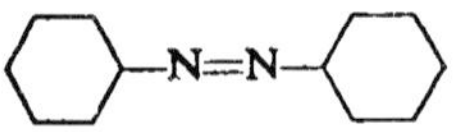

Skeleton formula for azo dyes

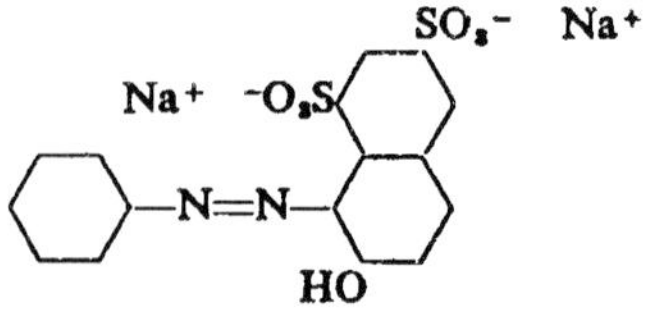

Orange G

The –N=N– group occurs once, twice, or thrice in the ion, which may be a cation or an anion; the corresponding group-names are mono-azo, disazo, and trisazo. The formula shows that orange G is an anionic, mono-azo dye; there are two–SO_3^- groups. The name refers to the colour, which is a very yellowish orange. This is a typical background dye.

The structural formula for Biebrich scarlet is shown here. It provides a good example of a highly conjugated disazo dye.

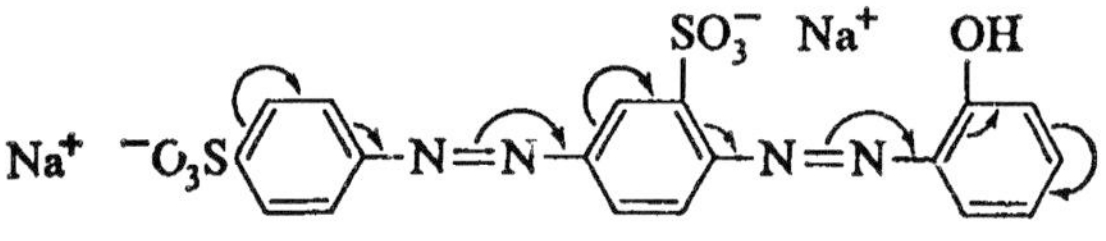

Biebrich scarlet

Not all azo substances are dyes. Sudan IV, for example, possesses the azo chromophore and is coloured (red). The formula

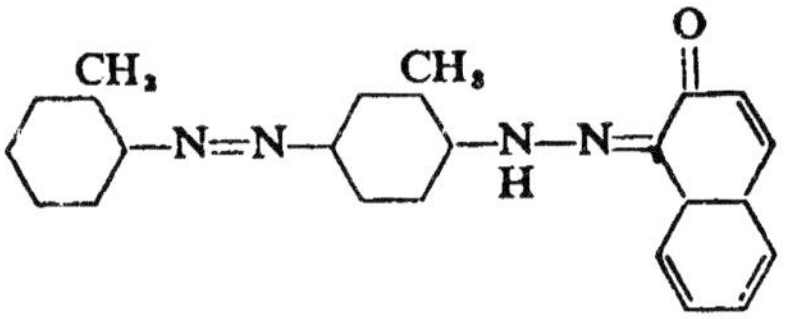

Sudan IV

shows, however, that it lacks an auxochrome. It is insoluble in water, but very soluble in lipids. It is an important lysochrome, used in histochemistry for the recognition of lipids. It is presented to the tissues at saturation in 70% ethanol. Being much more soluble in lipids than in this solvent, it accumulates in lipid globules and colours them. As we shall see in the next chapter, dyeing is a very different process from mere colouring by solution.

The Causes of Differential Dyeing

When a section is soaked in a solution of a single dye, the various tissue-constituents react differently. There is differential uptake, and the varying intensity of coloration serves to distinguish the constituents. The latter may also react differently to different dyes, for a particular object may take up much of one, little of another. These facts make it possible to produce striking colour contrasts in what had been a transparent, colourless object. That, indeed, is the purpose of dyeing. We are now concerned with the causes that produce these effects.

Whether a particular object in a microscopical section is strongly or weakly coloured by a particular dye depends on three factors: the *chemical affinity* between the object and the dye, the *density* of the object, and the *permeability* of the object by the dye.

Chemical affinity

Some of the constituents of the tissues, in the form in which they appear on the slide in sections, are acidic. Examples are DNA and chromatin, RNA and ribonucleoprotein, the matrix of cartilage, the mucous secretions of certain gland-cells, and certain conjugated lipids. Other constituents, such as the groundcytoplasm of most cells and the contractile substance of muscle, are neither particularly acidic nor particularly basic. They are markedly amphoteric; that is to say, their electric charge varies easily with the pH of the fluid in which they lie. Others again, such as collagen, the cytoplasm of red blood-corpuscles, and the granules of eosinophil leucocytes, are basic.

DNA, RNA, and phospholipids owe their acidity to their phosphoric groups. The acidity of the mucopolysaccharides of cartilage and of certain mucous secretions is due to sulphuric and carboxyl groups. The amphoteric proteins owe their character to a balance between acidic amino-acids (aspartic, glutamic, hydroxyglutamic, &c.) and basic ones (lysine, arginine, and histidine). The latter predominate in the basic proteins of collagen and red blood-corpuscles.

When a protein is put in a fluid at a pH below its iso-electric point, the amino-groups of its basic amino-acids tend to become

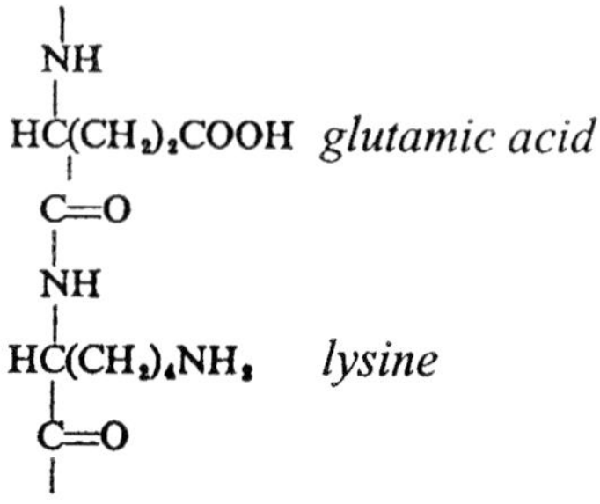

```
   |
   NH
   |
   HC(CH2)2COOH   glutamic acid
   |
   C=O
   |
   NH
   |
   HC(CH2)4NH2    lysine
   |
   C=O
   |
```

Part of a protein chain at the iso-electric point

positively charged as $-\overset{+}{N}H_3$. When the pH is above the iso-electric point, the carboxyl groups of the acidic amino-acids become ionized as $-COO^-$. If the proportions of basic and acidic amino-acids in a particular protein are about equal, the electric charge of the protein as a whole is easily shifted by variation in pH within the range of the fluids ordinarily used in microtechnique; but in fact all proteins can be shown to be amphoteric if there is a sufficient change of pH.

In chapter 6 we have seen that all dyes dissolve as ions, and that the coloured ions of some are positively charged, of others negatively. This can easily be proved by experiment. An agar gel is allowed to set in a vertical U-tube; and the dye-solution to be tested is added to both limbs of the tube, above the agar. The agar serves the mechanical function of holding the water in the U-tube nearly still. Salt must be dissolved with the agar, to permit the easy passage of an electric current. An electrode is placed in the dye-solution on each side. It is necessary to take the usual steps .o prevent polarization of the electrodes; full practical instructions for setting up the apparatus have been given elsewhere. When the current is switched on, the coloured ions begin to move down one or the other limb of the U-tube, towards either the cathode or the anode. The distance travelled in a given time depends on the concentration of the dye and of the agar, on the diffusibility of the dye, and on the strength of the electric current. In the apparatus referred to, a movement of 2 or 3 cm in 24 hours is usual. There is often a much smaller movement of the coloured ions down the other limb of the U-tube. This is due to simple diffusion.

Most dyes can be described unequivocally as cationic (= basic) or anionic (= acid). Some, however, are amphoteric. Haematein is an example. This dye is cationic below pH 6.6 (the iso-electric point) but anionic above.

In general, the acidic (negatively charged) objects in tissues attract the cationic dyes, while the basic (positively charged) objects attract the anionic dyes. Since the former objects are easily dyed by basic dyes, they are called basiphil; the latter, since they are easily dyed by acid dyes, are called acidophil. It is to be remembered that basiphil objects are acidic, acidophil objects basic. Thus the nucleic acids, most nucleoproteins (ordinary 'chromatin', for instance), the matrix of cartilage, many mucous secretions, and certain conjugated lipids are basiphil; while collagen, the cytoplasm of red blood-corpuscles, and the granules of eosinophil leucocytes are acidophil.

It is useful to form a concrete idea of the effects of cationic and anionic dyes by studying sections of specially selected tissues. The testis, intestine, and pancreas of the mouse are suitable. Most fixatives leave the tissues in a state in which cationic dyes act well, but some are unfavourable to anionic ones. It is well to use Zenker's fluid for the present purpose, because it is especially favourable to the action of anionic dyes. Sections should be lightly dyed and then run quickly through the ethanols and xylene into a resinous mounting medium.

In sections coloured by cationic dyes, the most prominent feature is the strong colouring of chromatin. Metaphase and anaphase chromosomes, both mitotic and meiotic, are more deeply coloured than anything else; nuclei are conspicuous on account of their chromatin content. The mucous secretion of the goblet-cells in the intestine may be rather strongly dyed, frequently in a different colour from anything else. The cytoplasm of most kinds of cells (such as spermatogonia and spermatocytes, and the interstitial cells of the testis) is scarcely tinged. Certain kinds of cells, however, have a large amount of ribonucleic acid in the cytoplasm, and this takes the dye. The epithelial cells of the crypts of Lieberkühn are examples; the bases of the exocrine cells of the pancreas are even more strongly dyed, because of the particularly great amount of ribonucleic acid associated with the highly developed ergastoplasm of this region. There is scarcely any colour, however, in the contractile substance of smooth muscle, in collagen fibres, in the free border of the intestinal epithelium, or in the large granules of the Paneth cells in the crypts of Lieberkühn. There is nothing to distinguish the granules of the eosinophil leucocytes in the cores of the intestinal villi, and these cells can scarcely be recognized. Red blood-corpuscles are not dyed.

The whole aspect of sections of the same tissues coloured with a typical anionic dye is entirely different. There is a general colouring of the cytoplasm of most kinds of cells. The cytoplasm of the interstitial cells of the testis takes up more of the dye than that of the spermatogenetic cells, and groups of the former kind of cells therefore stand out rather prominently under the low power of the microscope. The basal part of the exocrine cells of the pancreas is moderately deeply coloured. The cytoplasm of the epithelium of the crypts is pale. The contractile substance of smooth muscle takes the dye; so do collagen fibres. None of these tissue-constituents, however, is nearly so strongly coloured as certain cytoplasmic granules namely those of the Paneth cells and particularly the minute ones of the eosinophil leucocytes in the cores of the villi. Chromatin shows no special affinity for the dye and is only lightly coloured. The nuclear sap is colourless. The nuclei therefore often appear almost as though they were cavities in the dyed cytoplasm, and they are not nearly so conspicuous as in sections coloured with a cationic dye. Red blood-corpuscles will take any anionic dye that is able to enter them.

It must not be supposed that basiphil objects are necessarily resistant to coloration by anionic dyes, or acidophil ones to coloration by cationic dyes. Proteins that are predominantly acidic contain some basic amino-acids, and conversely. Thus all the proteins in a cell can eventually be dyed to some extent by any dye, whether basic or acid. However, by controlling the times during which the dyes act, it is easy to ensure that the basiphil objects will be dyed by cationic (basic) dyes and the acidophil ones by anionic (acid) dyes, each almost to the exclusion of the other.

The amphoteric objects in cells can be dyed with either cationic or anionic dyes, by controlling the pH. In practice one usually wants to leave the cytoplasm uncoloured, or else to dye it differently from the chromatin. If a cationic dye is used in slightly acid solution (methyl green, for instance, in a weak solution of acetic acid), the chromatin will be coloured: but the amphoteric cytoplasm, having been rendered basic by the acidity of the solution, will have very little affinity for the cationic dye, and the cytoplasm will therefore appear colourless or nearly so. Alternatively one may use a cationic dye in acid solution first, and then an anionic one of contrasting colour. The amphoteric cytoplasm will have some affinity for the anionic dye, and it can thus be coloured differentially.

Since most tissue-constituents will eventually be coloured to some extent by any dye, it may be desirable to limit the period of dyeing, so that some of them will appear colourless when others have already been quite strongly dyed. In other words, dyeing may be arrested at a particular stage instead of being allowed to go to completion. This is *progressive* dyeing, so called because the depth of colour becomes progressively greater until the desired end is reached. In *regressive* dyeing, on the contrary, the tissues are allowed to take up an excess of the dye, and the latter is then partly removed, with careful control under the microscope, until certain tissue-constituents have lost the colour while others retain it. During this regressive process the parts become more distinct from one another than they were when all parts held too much of the dye. The regressive stage of the process is therefore called *differentiation.*

Cationic dyes are frequently used regressively. It is customary to use acids as differentiating agents. They act, in principle, in exactly the same way as when they are mixed with the dye: reliance is placed on the ability of the acid to render the amphoteric tissue-constituents basic, and thus make them let go of the cationic (basic) dye. Alkalis can be used in the same way to differentiate acid dyes.

An alternative way of differentiating is to soak the tissue in a fluid in which the tendency of the dye to remain ionized is reduced or eliminated, but in which the dye is soluble. Ethanol, either absolute or in strong aqueous solution, is often used for this purpose. The dye is rapidly lost from those tissue-constituents that hold little of it. If differentiation be stopped as soon as these constituents appear colourless, the objects that still retain a considerable amount of the dye will contrast strongly with them. The tissue must now be brought quickly into a fluid, such as xylene, in which the dye is insoluble.

There is sometimes evidence that a dye is held to a tissue-constituent not by its electrostatic charge, but by a hydrogen bond. It will be remembered that one speaks of this bond when two atoms are linked together by a hydrogen atom, which acts almost as though it were bivalent. The atoms that can be bound together in this way by an intervening hydrogen atom are the strongly electronegative ones, oxygen, nitrogen, and fluorine (though the latter is not important from the point of view of microtechnique). A hydroxyl group and an amino-group, for instance, can be bound together by this bond. Since two electronegative atoms are concerned in such cases (oxygen and nitrogen), it is evident that we are not dealing here with a reaction between positively and negatively charged atoms.

Hydrogen bonding is to be regarded as probable in these circumstances

(1) if the action of the dye is insensitive to pH;

(2) if the action of the dye is insensitive to an electrovalent competitor such as a 10% sodium chloride solution;

(3) if saturating the dye solution with urea prevents dyeing (since urea forms hydrogen bonds readily and occupies the available sites to the exclusion of the dye);

(4) if the action of a cationic dye is unaffected when the

carboxyl-groups of the tissue-constituents have been blocked by methylation (i.e. the hydrogen atom has been replaced by $-CH_3$);

(5) if the action of an anionic dye is unaffected by the destruction of the amino-groups in the tissue by nitrous acid.

One of the tissue-constituents that is most obviously dyed by hydrogen bonding is the elastin of elastic fibres. There are several complex dyes, of uncertain chemical composition, that are more or less specific for this substance. Resorcin-fuchsine is one of these. The kinds of evidence listed above show that they attach themselves to elastin by hydrogen bonds, but it is not possible to say which oxygen and nitrogen atoms are involved in the bond.

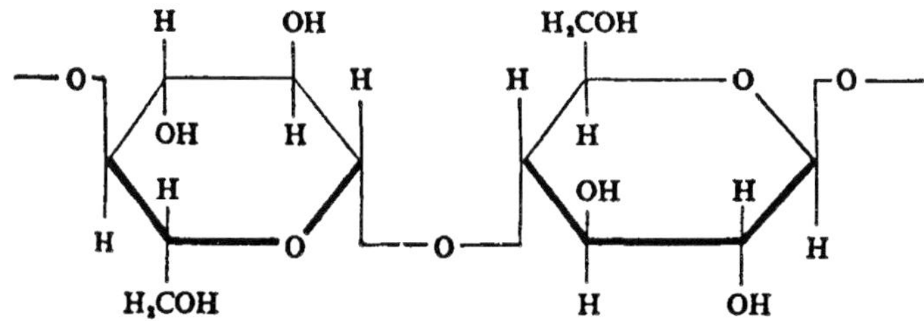

Part of a cellulose molecule

Acidic tissue-constituents, lacking any basic groups, can in certain circumstances be dyed by anionic dyes. Thus methyl blue, an anionic triarylmethane dye, will colour cellulose cellwalls. Indeed, it is used in the textile industry for dyeing cotton, and is often called cotton blue. The structural formula for cellulose shows the absence of any group that would be thought likely to attract an anionic dye. It is supposed that the linkage is through a hydrogen bond, which may tie the oxygen of a hydroxyl group in cellulose to the nitrogen of an amino-group (or substituted amino-group) in a dye. Methyl blue, and certain other anionic triarylmethane dyes that will colour cellulose, contain nitrogen in a substituted amino-group.

The peptide groups $\left(\begin{matrix}-\underset{\underset{O}{\|}}{C}-\underset{H}{N}-\end{matrix}\right)$ of nylon are thought to form hydrogen bonds with the amino-groups of certain dyes. If so, it is likely that the peptide groups of protein could behave similarly. This may account for the tendency of certain acid dyes to colour tissue-constituents without appearing to discriminate between those that are positively and negatively charged.

It is thought by some that dyes are first brought close to particular tissue-constituents by the interaction of electrical charges, and are then tied closely to them by 'short-range' forces; that is to say, by covalent or hydrogen bonds.

Workers in the textile industry have made important advances in the understanding of chemical affinity for dyes by undertaking quantitative studies, but this approach is necessarily much more difficult under the conditions of dyeing in microtechnique, where everything is on a minute scale and the substrates that attract the dyes are intimately mixed together. Nevertheless, a promising start has already been made in the application of quantitative methods to this field. The work has so far been confined almost entirely to a single cationic thiazine dye.

It has already been remarked that dye-ions are usually taken up without change of colour. An important exception to this rule must now be briefly mentioned.

Certain pure (unmixed) dyes have the property of dyeing particular tissue-constituents in a colour that differs from that of the solution. This is called *metachromasy* and the dye is called *metachromatic;* the tissue-constituents that are dyed in a differing colour are called *chromotropes* (colour-turners).

All the metachromatic dyes that are commonly used in microtechnique are basic, and they all act on the same chromotropes.

The metachromasy of acid dyes will not be considered in this book.

The metachromatic dyes do not form a chemical group by themselves. On the contrary, they are distributed among all the major groups except the azo. Methyl violet is a metachromatic triarylmethane dye, for instance; thionine and azure B are metachromatic thiazines. (Commercial thionine is often adulterated with a red dye, and may then give false metachromatic colours.) It is a curious fact that when all the hydrogens of the aminogroups of a dye are replaced by methyl, the resulting substance is not metachromatic. Thus crystal violet is not metachromatic (though in commerce it is often adulterated with the strongly metachromatic methyl violet).

The colour-shift caused by chromotropes is always in the same direction. Green dyes become bluish; blue dyes are reddened to purple or red; red ones become orange or yellow. This means that the absorption-maximum of the dye moves in all cases towards the shorter wave-lengths.

Chromotropes are necessarily acidic, since they have affinity for cationic dyes. Their negative charges are due to the possession of sulphuric, phosphoric, or carboxyl groups. Many of the most familiar chromotropes are sulphuric esters of polysaccharides of high molecular weight. These are often mucosubstances. Thus the matrix of cartilage, the secretions of certain mucous glands, and the granules of the basiphil cells *(Mastzellen)* of connective tissue give metachromatic colours because they contain chondroitic acid, mucoitic acid, and heparin respectively. Not all chromotropes, however, are mucosubstances. Agar, strongly chromotropic, lacks an amino-group.

The cause of metachromasy has not been established with certainty. The metachromatic dyes have a tendency to form dimers and polymers, of the metachromatic colour, in aqueous solution. Chromotropes appear to be substances that favour strongly the dimeric and polymeric forms of the dye, and hold the dye to themselves in its dimeric or particularly its polymeric form. A small molecule would not be able to hold a high polymer of a dye, but a polymeric substance, with numerous negative charges arranged on its surface at regular and convenient distances apart, might be able to do so. It thus appears that the mode of attachment of the dye causes the shift of colour.

Metachromasy is important in cytological technique because it acts as a pointer towards chemical composition.

DENSITY

Make two gelatine gels, one containing 5% and the other 20% of gelatine. Fix these in formaldehyde solution (or in a mixture containing formaldehyde). Cut a section of each on the freezing microtome, at the same thickness (say 15 μ). Allow an aqueous solution of any dye to act on both sections for the same period, which must be sufficient to allow complete permeation. Rinse with distilled water. It is not surprising that one section is more strongly coloured than the other. The result is due solely to the fact that one section is *denser* than the other: that is to say, it contains more colourable matter than the other, and therefore takes up more dye.

One often reads that the cytoplasm of a particular cell is 'dense'. How did the writer know? The depth of colouring may be due to quite different causes. If one had not made the gelatine gels oneself, one would not have known the cause of the difference in colouring. The density of the two sections might have been exactly the same, but one of them might have been made of a basic and the other of an acidic protein, and they would necessarily have taken up different amounts of any particular dye.

Anyone who performs the experiment just described is likely to be surprised by the darkness of the section that contains more gelatine. There is only four times as much gelatine, but it appears to have taken up much more than four times as much dye. This is, in fact, an illusion. Let us suppose that we have used a black dye, and that the section made from the weaker gelatine gel stops one-half of the light that is shone on to its surface. Make a pile of four such sections, one on top of the other, and illuminate them from below. The second from the bottom receives half the original light, and sends up half of what it receives to the next; this halves the light again, and so does the uppermost section. Thus only one-sixteenth of the light gets through. But in these four sections together there is the same amount of gelatine-and therefore the same amount of dye-as in a single section cut from the stronger gel. The latter, for the same reason, lets through only one-sixteenth of the incident light, though the amount of dye is only four times as great as in the section that let through one-half. For this reason we are apt to form an exaggerated opinion of the difference in the amount of dye taken up by objects in microscopical preparations.

When dividing cells are coloured with a cationic (basic) dye, each metaphase chromosome takes up far more of it than an equal volume of cytoplasm does. This results partly from the fact that nucleoprotein is much more acidic (basiphil) than the cytoplasm, but partly also from the fact that the chromosomes are denser. Their density is indicated by the fact that *anionic (acid)* dyes also colour chromosomes more deeply than the cytoplasm; indeed, they are used to colour chromatin in several familiar techniques. For instance, acid fuchsine is used for this purpose in Mallory's method for the differential colouring of collagen. [114] Whenever a particular object can be more deeply dyed than the cytoplasm or nuclear sap by both anionic and cationic dyes, it is likely to be denser (though one requires an interference microscope for the actual measurement of density). The converse, however, is not necessarily true: we cannot conclude that a particular object contains little matter, from the fact that it takes up little or no dye. It may possess few or no acidic or basic groups, capable of attachment to cationic or anionic dyes.

It may be mentioned that certain anionic dyes, particularly methyl blue, have a strong affinity for chromatin. This property has been carefully investigated, but a completely satisfactory explanation has not been found. The dye attaches itself to DNA.

PERMEABILITY

The amount of dye taken up by an object in a particular time does not depend solely on the amount of matter it contains, or on the abundance of charged groups. It depends also on whether the dye can permeate the object easily. Certain dyes have a great capacity for penetration, others very little. It is easy to test this capacity. It is only necessary to allow some gelatine or agar to set into gels in test-tubes, and then to add solutions of different dyes, all at the same concentrations, to different tubes. Some dyes diffuse quickly into the gel, some slowly. Eosin (xanthene) and orange G (azo) are examples of rapidly permeating dyes, methyl blue (triarylmethane) of slowly permeating. In general, the rapidly permeating dissolve as single ions (and not very large ones) ; the slowly permeating tend to form colloidal solutions, in which each particle is an aggregate of several ions.

In microtechnique we make use of the varying rates of penetration to colour differentially objects that carry the same electric charge. Thus collagen and the red blood-corpuscles of vertebrates both contain a high proportion of basic protein and are therefore acidophil. It is quite easy, however, to colour them differently with two different anionic dyes. The mixture of methyl blue and eosin devised by the Oxford histologist, Mann, serves the purpose well. Methyl blue is much the more powerful dye (partly, perhaps, because a single positive charge in the object may be able to attract a whole ion-aggregate). Wherever the two dyes can compete, methyl blue predominates and the colour is blue or bluish. This applies to collagen fibres, which are evidently loose-textured, for they are easily coloured by aggregated dye-ions. Red blood-corpuscles, on the contrary, are very close-textured, and they are much more easily entered by separate dye-ions than by aggregates. The result is that Mann's methyl blue/eosin dyes collagen blue and red blood-corpuscles orange-red.

The various tissue-constituents can be arranged in the order of their permeability. Collagen is very, permeable; cytoplasm less so; the contractile substance of muscle less so again: red blood-corpuscles are particularly impermeable. It is possible to dye collagen in one colour, cytoplasm and contractile substance in another, and red blood-corpuscles in a third, by the use of three different anionic dyes. The result is achieved without any reliance on different chemical affinities.

The three qualities in the tissue-constituents that make possible the differential action of dyes-chemical affinity, density, and permeability-may act together or may antagonize one another. Chromatin, for instance, is chemically reactive (basiphil), dense, but permeable: all these characters act together to render it easily coloured by cationic dyes. Red blood-corpuscles, on the contrary, are chemically reactive (acidophil), dense, but very impermeable. A microscopical preparation coloured with two or three dyes owes its variegated appearance to complex interactions, and it is not always easy to disentangle the effects of the three factors.

Those anionic dyes that penetrate readily tend to colour all acidophil and amphoteric tissue-constituents about equally deeply, and they are therefore diffuse in action. They are useful 'backgr-ound' dyes, giving colour-contrast to a cationic dye used for chro-matin. If methyl blue or a similar anionic dye were used instead of a diffuse one, it might spoil the effect by giving some of its colour to chromatin.

It is best to choose background dyes of colours near the mid-dle of the visible spectrum (yellowish or green), since these ap-pear to the human eye as 'unsaturated'; that is to say, they stimu-late all the colour-receptors in the retina to some extent and thus appear pale, as though mixed with white. Orange G, being yel-low, is suitable. For contrast the chromatin should be dyed in a 'saturated' colour near one end of the spectrum (red, blue, or vio-let). Violet gives the best contrast with yellow, because the col-ours are complementary.

If chromatin be dyed black or grey, any background dye will necessarily reduce the contrast.

THE ACTION OF MORDANTS

Up till now we have been concerned with the direct attachment of dyes to tissue-constituents. In some of the most important pro-cesses of dyeing, however, an intermediary or *mordant* stands be-tween the dye and the tissue. The dye attaches itself to the mor-dant: the latter (as its name suggests) 'eats into' or grips the tis-sue.

The great advantage of mordanting is that the colour is not re-movable by neutral fluids, whether aqueous or alcoholic. One may therefore colour progressively or regressively until the tissue is properly displayed, and then dehydrate at leisure, or counterstain as desired. The fact that dehydration may be done as slowly as one pleases is particularly helpful in making whole mounts. As we shall see, it is chiefly chromatin that is coloured by most mordant dyes. Nuclei are crowded together in the epithelial parts of many organs, but are sparse in the connective tissues; and there may be cavities from which they are absent. Thus the dyeing of chromatin gives clear micro-anatomical pictures in whole mounts, even with quite low powers of the microscope.

The fastness of mordant dyes makes them suitable for use in aqueous mounts, provided that acidity is avoided.

All dyes that act with mordants can also be used without them, but if so their effects are quite different. By no means all dyes can be used with mordants. The chief ones so used in microtechnique are carminic acid and haematein. The chief mordants used with them are salts of aluminium and of ferric iron.

In the textile industry the chief dyes used with mordants are azo dyes, and the chief mordants are complex basic salts of chromium. The chemistry of the mordanting of azo dyes by chromium has been carefully worked out by the textile chemists. Unfortunately it is radically different from the mordanting of carminic acid or haematein by aluminium or ferric salts.

One of the simplest dyes that can be mordanted is purpurine, which, like carminic acid, is an anthraquinone dye. Alizarine is even simpler, but its low solubility in suitable solvents makes it less convenient in practical use. The third –OH group, which

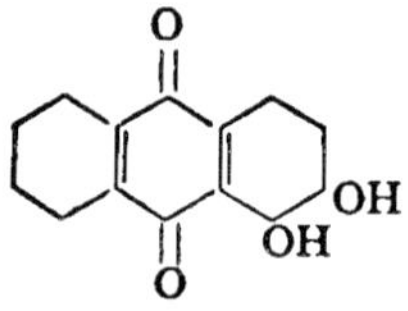

Alizarine 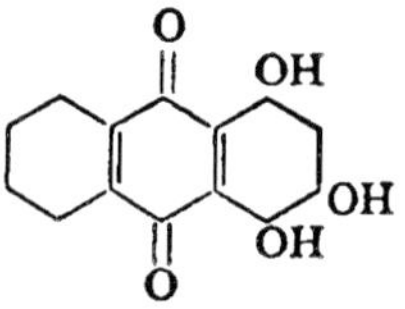 *Purpurine*

occurs in purpurine but not in alizarine, is rather unreactive, and the two dyes tend to behave similarly with mordants. Both dyes occur naturally in the form of glucosides in the root of the madder plant, *Rubia tinctorum* (Rubiaceae), but the synthetic products are almost invariably used. Purpurine is unfortunately named, for it is a red dye.

If purpurine is dissolved at saturation in 60% ethanol, it acts as a typical though very weak acid dye. If, however, the dye is dissolved in a solution of aluminium sulphate, the result is entirely different. The following is a convenient solution.

Take 0.8 g of purpurine and 7.88 g of aluminium sulphate crystals ($16H_2O$); add 450 ml of 60% ethanol; boil with reflux condenser until the solids have dissolved; cool; make up to 500 ml with 60% ethanol.

The purpurine in this solution, if pure, is at M/160, the aluminium sulphate at M/40. The solution may be called standard aluminium purpurine, or 'purpural' for short.

The ethanol in this solution increases the solubility of the dye and thus makes the solution last longer when used repeatedly; it slows down or prevents the growth of bacteria and moulds; and it decreases ionization and thus 'equalizes' the action on tissues (that is to say, prevents the solution from over-dyeing locally).

Restrainers such as ethanol and glycerol are particularly useful in the dyeing of whole mounts, but they sometimes serve a useful purpose with sections also, especially when dyes are used progressively.

The various tissue-constituents are coloured by purpural exactly as though it were a basic dye. There is, however, this difference, that whereas basic dyes can be washed out of the tissues by neutral alcoholic solutions (50% or 70% ethanol, for instance), mordanted purpurine cannot.

Other soluble salts of aluminium (the chloride, for instance) may be substituted for the sulphate: the action is again that of a basic dye. If, however, sodium sulphate is used instead of the aluminium salt, there is no mordanting and purpurine acts as a very weak acid dye. It is the cation of the salt that mordants.

It will be remembered that metallic aluminium has three electrons in its outer 'shell' (the third or 'M' shell). In its salts these are lost to the anion, and the cation is left with the same orbital electrons as the inert gas neon. The cation has, however, the ability to accept electrons from suitable donor atoms, and in the presence of water it accepts no fewer than 12, two from each of 6 oxygen in water molecules. In solution the cation consists of aluminium in the centre, with 6 molecules of water surrounding it. The 6 dative covalency bonds are arranged in opposite pairs, each pair being at right-angles to the other two pairs. Each valency bond holds a molecule of water.

It may be wondered why exactly 6 bonds are formed, involving the donation by water of 12 electrons. This is controlled partly by the geometrical fact that 6 water molecules fit round a central atom or ion more easily than (say) 5 or 7, but there is also another cause. The reader may recall that the M shell will hold a total of 18 electrons, but these cannot all be accepted when the atomic nucleus is that of aluminium. Twelve have the faculty of 'hybridizing' with one another, without distinction as to whether they belong to the *s*, *p*, or *d* orbitals of the shell; and these 12 give the dative covalency of six.

The complex water/aluminium ion, having 3 electrons less than the total number required to neutralize the protons of its atomic

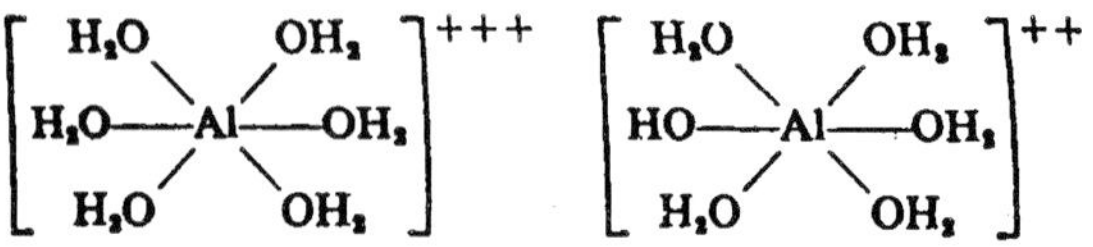

$$\left[\begin{array}{ccc} H_2O & & OH_2 \\ & \diagdown \;\; \diagup & \\ H_2O - & Al & - OH_2 \\ & \diagup \;\; \diagdown & \\ H_2O & & OH_2 \end{array}\right]^{+++} \qquad \left[\begin{array}{ccc} H_2O & & OH_2 \\ & \diagdown \;\; \diagup & \\ HO - & Al & - OH_2 \\ & \diagup \;\; \diagdown & \\ H_2O & & OH_2 \end{array}\right]^{++}$$

Structural formulae for hydrated aluminium ions. The one on the right has lost a hydrogen ion.

nuclei (aluminium, oxygen, and hydrogen, together), may be regarded as holding 3 positive charges at the moment of its formation. The 6 water molecules, however, do not all remain intact. One of them, at least, tends to lose a hydrogen ion, which joins a molecule of the solvent water to form a hydronium ion, H_3O^+, and thus departs. Since it carries off a positive charge, the complex ion is only doubly charged; but the solution, having gained a hydronium ion, has become acid. Thus the pH of an M/40 solution of aluminium sulphate in 60% ethanol is 3.08.

When purpurine is dissolved in water or in solutions of ethanol in water, it behaves like a phenol; that is to say, like a weak acid. A saturated solution in 60 % ethanol has a pH of 5.20. Some of the hydrogens of the –OH groups are ionized and lost, and the dye is now negatively charged. In the structural formula shown

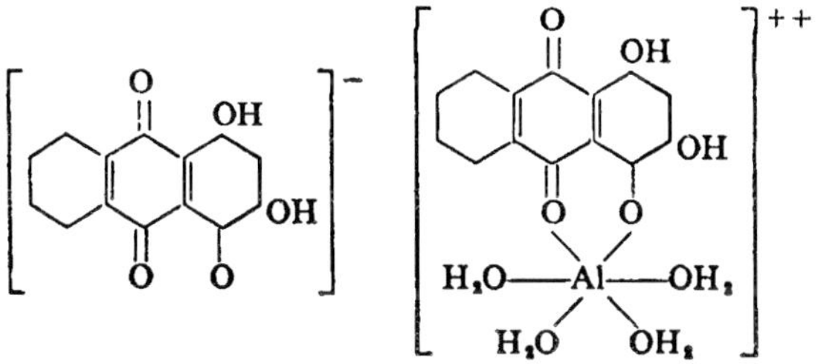

Ionized purpurine *The aluminium purpurine ion*

here, one of the –OH groups is ionized. There are now two oxygen atoms near one another, both capable of donating electrons to suitable acceptors. If purpurine is added to a solution of an aluminium salt, these two oxygen atoms will tend to replace two of the water molecules that are combined with aluminium. Thus the purpurine will form a complex or 'co-ordination compound' with the aluminium.

The purpurine ion might attach itself either to an aluminium ion that was associated with 6 water molecules, or else to one that was associated with 5 water molecules and one -OH group (top right); but the compound would probably be the same in both cases, for the acceptance of purpurine by the aluminium atom would be likely to be accompanied by the acceptance of a hydrogen ion from a hydronium ion in the solution, which would attach itself to the –OH group of the hydrated aluminium ion.

Since the fully hydrated aluminium ion has a triple positive charge, while the purpurine ion brings with it a single negative charge, the complex ion as a whole has a positive charge of two.

In the practical dyeing solution, standard aluminium purpurine there is only one molecule of purpurine to 8 atoms of aluminium. The composition of the solution may be expressed in terms of the *mordant quotient;* that is to say, the number of atoms of mordant metal in the solution, divided by the number of dye molecules. In the standard aluminium purpurine solution, then, the quotient is 8. It must therefore be supposed that the majority of the hydrated aluminium ions in the solution that are combined with purpurine at all are combined with only one ion of it.

When ionized purpurine has associated itself with aluminium, a new ring has been formed, consisting of 6 atoms arranged in the following order: aluminium, oxygen, carbon, carbon, carbon, oxygen. This ring is particularly stable. Since the purpurine sends out two bonds that grip the aluminium like the chela of a lobster, the compound is said to be chelate.

Since the aluminium-purpurine ion in the dye-solution is positively charged, it will tend to combine with basiphil (negatively charged) tissue-constituents, such as the nucleic acids. It will not tend to combine with acidophil (positively charged) constituents, such as the haemoglobin of red blood-corpuscles.

It might be thought that the hydrated aluminium ion, carrying a triple positive charge, would have a greater tendency to be attracted to sites of negative charge in the tissue than would the doubly charged aluminium-purpurine ion. If so, little dyeing would be likely to occur; yet the solution dyes strongly. The explanation appears to be as follows. When some of the water molecules of a coordinaton complex, such as the hydrgated aluminium ion, are replaced by other molecules, the solubility of the complex in water is reduced. This reductin in solubility will favour the deposition of the aluminium-purpurine ion on the sites of negative charge, while the hydrated aluminium ion has a strong tendency to remain dissolved.

A few of th hydrated aluminium ions are likely to take up two purpurine ions; some may take up three. The result will be a reduction of the positive charge respectively to one or nought; in the latter case there would be precipitation of an insoluble pigment. On the analogy of the compound of purpurine with cobalt, it must be regarded as probable that one aluminium atom can accept chelate bonds from three molecules of purpurine. This, however, is not certain. A great deal of study has been devoted to the attempt to discover the exact composition of the flaming red dye, Turkey red, which has been known in the East for many hundreds of years. It appears certain that although this substance contains calcium, yet the chelate bonds are all with aluminium, and that only *two* alizarine molecules are thus joined to each aluminium atom: of the two remaining covalencies of this metal, one still retains a water molecule (and the other is linked through oxygen to calcium).

Compounds between mordants and dyes are known as *lakes*. The artists' water colour, madder lake, is of this nature. This insoluble substance is similar in composition to Turkey red. It is distributed by the artist's brush in the form of a fine suspension. When a mordant and a dye are disslved together, a lake may remain in solution, but there is often a tendency towards general preciptation of an insoluble lake, in which the mordant metal has combined with the maximum number of dye ions.

Although electrostatic attraction presumably draws the aluminium-purpurine ions towards sites of negative charge in the tissues, yet a bond other than the ionic must eventually be established, for otherwise the mordant/dye complex would be no more tightly held to the tissues than basic dyes are, and like the latter it would be washed away readily enough by neutral ethanol. It is probable that covalent linkages are established between the aluminium atom and acid groups in the tissues, such as the phosphoric group of the nucleic acids. One or two of the four water molecules remaining attached to the aluminium atom would be set free, and one or two covalent linkages formed with the acid group in the tissue. The positive charges on the mordant/dye complex would be reduced to one or nought.

Similar linkages are likely to be formed with carboxyl and other acid groups in proteins and other tissue-constituents.

It is not always necessary to prepare a mordant/dye complex first and subsequently to allow it to come into contact with the tissues. On the contrary, one may first soak the tissues in a solution of the mordant and then in a solution of the dye. The basiphil tissue-constituents will take up the positively charged hydrated aluminium ion and the dye will displace water from the attached aluminium complex when given the chance to do so. It might, indeed, be said that the dye is being used as a reagent for the detection of aluminium (though one might be misled if the tissues happened to contain another metal capable of mordanting the dye). Both the single-bath method (in which the dye is mixed with the mordant) and the two-bath method (in which the tissue is put first in the mordant and then in the dye) are of common use in microtechnique.

When the single-bath method is used, one may dye either regressively or progressively, but regressive dyeing is almost invariable with the two-bath process. There are two chief ways of differentiating mordant dyes after deliberate over-dyeing. One may extract the excess of dye by placing the tissue either in an acid (often a weak solution of hydrochloric acid) or else in a solution of the substance that was used to mordant.

It always surprises students to learn that the mordant-the very substance that attaches the dye to the tissue—can also be used to remove the dye from it. The fact is that everything depends on the relative abundance of mordant and dye. When a section has been dyed, the total amount of dye in it is very small. If the section is now placed in a solution of the mordant, the latter is present in enormous excess. The dissolved aluminium competes with the attached aluminium for the minute amount of purpurine that is colouring the tissue. The dye redistributes itself. If sufficient time be allowed, no visible trace of dye will remain in the tissue. In practice one looks at the section under the microscope from time to time. When the objects that it is desired to show clearly are still strongly coloured, but the surrounding parts of the tissue have become colourless or scarcely tinged, one washes away the differentiating fluid with water.

While the dye is being extracted by the mordant, coloured clouds may be seen in the solution surrounding the section. Thus the solution *extracts* the dye, though both mordant and dye are present in it. It follows that whether a mordant/dye solution dyes or extracts depends on the relative abundance of dye and mordant; that is to say, on the mordant quotient. For any particular mordant and dye one can find a concentration of each that neither increases nor decreases the intensity of colouring of tissue-constituents that have already been dyed by the same mordant and dye. In experiments of this sort it is best to work with sections that have been exposed to the standard aluminium purpurine solution until chromatin has been strongly coloured but cytoplasm only feebly. The mordant quotient of an aluminium purpurine solution that neither increases nor decreases the intensity of colouring of such a section is called the *critical mordant quotient.* [23] If the quotient of a solution is lower than the critical figure, the solution acts as a dye; if higher, as a differentiating agent. In the case of aluminium purpurine, with the aluminium salt dissolved at M/40 in 60% ethanol, the critical quotient is 16.

If an *undyed* section be placed in an aluminium purpurine solution in which the mordant quotient is at the critical figure, dyeing will take place very slowly until chromatin is strongly and cytoplasm feebly coloured; no further increase of colour will occur, however long the section may remain in the dye.

Solutions in which the mordant quotient is not very many times less than the critical figure are particularly easy to use, because there is not much tendency to over-dye. Standard aluminium purpurine ('purpural'), with a mordant quotient of 8, is an example. It can be recommended to beginners as a routine dye for chromatin and other negatively-charged tissue constituents. It is to be used progressively. The period of dyeing varies according to the fixative used, but half-an-hour or an hour usually suffices. The section is then simply washed in 50% or 70% ethanol.

It will be remembered that carminic acid, like purpurine, is an anthraquinonoid dye. It possesses an =O and a phenolic –OH placed in the same relation to each other as in purpurine. It behaves towards mordants very much like the latter dye. Its colour (crimson) does not undergo much change when it links itself to aluminium, the mordant most commonly used.

One of the best solutions is Mayer's carmalum, which contains carminic acid, potassium alum, and water, with thymol, salicylic acid, or sodium salicylate as disinfectant. It is suitable for use with both sections and whole mounts. It may be used either progressively or regressively. Potassium alum (5% aqueous solution) readily extracts any excess of colour. This is a very easy dye to use, since it lends itself to leisurely progressive colouring. The colour is well maintained in both hydrophil and hydrophobe mounting media.

Haematein is perhaps the most valuable of all dyes used in micro technique. By itself it is an anionic dye, giving a very pale yellowish colour to basic tissue-constituents, but possessing no advantage over other anionic dyes; it possesses an =O and a phenolic –OH, so situated in relation to one another as to provide means of chelation with mordants. The chief mordants are aluminium sulphate, $Al_2(SO_4)_3$ (usually presented as potassium or ammonium alum), and ferric sulphate, $Fe_2(SO_4)_3$ (presented as 'iron alum', the ferric ammonium salt). When mordanted by either of these, haematein acts as a cationic dye. It gives a blue colour with aluminium, black or blue-black with ferric iron.

Aluminium haematein is chiefly used as a routine dye for chromatin in histology, in the mixtures of Delafield, Ehrlich, and Mayer. Haematoxylin is generally used instead of haematein in making up dye solutions. It will be remembered, that this is a colourless substance, converted to the dyehaematein, by oxidation. Haematoxylin is 100 times as soluble in ethanol as haematein, and this is probably why the former is chosen. Reliance is often placed on gradual oxidation by atmospheric oxygen.

A convenient substitute for the various aluminium 'haematoxylins' can easily be made with ethylene glycol as solvent for haematein, for the dye is quite soluble in this fluid and in mixtures of it with water. The dye solution is so easy to prepare and use that full instructions will be given here.

Prepare the following stock solutions. In practical microtechnique it suffices to weigh the solids to the nearest decigram.

ALUMINIUM SULPHATE, $\frac{M}{40}$ AQ.

$Al_2(SO_4)_3.16H_2O$	15.76 g
distilled water	to make 1 litre

HAEMATEIN, $\frac{M}{160}$ IN ETHYLENE GLYCOL, 50% AQ.

haematein	1.876 g
ethylene glycol, 50% v/v aq. to make 1 litre	

The stock solutions are ready for use as soon as they have been made.

The dye, 'haematal 8', is prepared by mixing equal volumes of the two stock solutions, and is ready for instant use. Haematal means *haematein* with aluminium. The solution contains 8 atoms of aluminium to each molecule of haematein, if pure substances were used in making up the solutions.

The dye works well after any routine micro-anatomical or histological fixative. Short fixation (3 to 6 hours) with Zenker's fluid is perhaps the method of choice.

Dyeing is carried out as follows.

1. Bring the slide to water.
2. Dye progressively. (The necessary period is often about 10 minutes.)
3. Wash in running water for about 3 minutes.
4. Rinse with distilled water.
5. Counterstain with a 0.1% aqueous solution of Biebrich scarlet for about ½ minute. (Eosin is also suitable.)
6. Rinse with distilled water.
7. Dehydrate with the ethanols.
8. Pass through xylene into DPX or Canada balsam.

Chromatin and other strongly basiphil substances are dyed blue; acidophil substances pink.

Haematal 8 gives results similar to those produced by the various aluminium 'haematoxylins', but it is rational, not empirical, in composition, is ready for use directly it has been prepared, acts quickly, and requires no differentiation.

Ferric haematein attaches itself more securely to the tissues than aluminium haematein does, and in particular is less easily removed by acids. It does not act quite like an ordinary cationic dye, for it attaches itself not only to strongly basiphil objects, but also to mitochondria, centrioles, and lipochondria. The difference between the link formed with the tissue-constituents and that formed by aluminium haematein has not been satisfactorily explained.

When ferric salts and haematein are dissolved together, an insoluble lake tends to be deposited. For this reason the two-bath method is generally used. The section is first soaked in iron alum or some other ferric salt and then transferred to a simple solution of 'ripened' (partly oxidized) haematoxylin. The latter is usually dissolved in ethanol, in which it is very soluble, and the solution subsequently diluted with water. The regressive method is used.

The excess of the dye is removed by soaking the section a second time in the mordant.

Heidenhain's method is the best-known example of this way of using iron haematein. Experienced workers can get precisely the result they require, but beginners find the differentiation rather difficult, because the extraction by the mordant seems to go faster and faster towards the end, just when one would like it to go slowly. Heidenhain's is one of the most important of all methods of dyeing in microtechnique. Almost everything in the cell can be revealed by careful differentiation, if the tissue was appropriately fixed. The black or blue-black colour gives excellent images in the microscope and is very convenient for photomicrography. Preparations are permanent in hydrophobe mounting media, even in those (such as Canada balsam) that tend to bleach certain dyes. It is generally best not to use an acid dye in addition, because the contrast between black objects and their surroundings is necessarily reduced by any background colour.

When using ferric haematein one need not wait for the 'ripening' of haematoxylin by atmospheric oxygen, nor need an oxidizing agent be added. Once again one can rely on the solubility of haematein in ethylene glycol. A convenient solution, ready for immediate use, may be prepared as follows.

Make a 10% v/v aqueous solution of ethylene glycol. Put 100 ml of this in a flask and add 0.08 g of haematein. Bring to the boil; simmer for 5 minutes; cool thoroughly; make up to 100 ml with distilled water. 02 g of sodium p-hydroxybenzoate may be added as a disinfectant.

The solution is to be used in the same way as Heidenhain's haematoxylin.

It is possible (though not usually helpful) to use the two-bath method with aluminium haematein. The mordanted dye may in this case be differentiated readily by acid. A modification of this method provides an easy proof of the fact that when a mordanted dye is differentiated by acid, the tissue/mordant link is attacked. It is only necessary to take two slides, treat both of them with the mordant (aluminium sulphate), and then put only one of the slides into the acid used for differentiation (weak sulphuric acid). After the acid has been washed away, both slides are put in the dye for the same length of time. The section that has been in the acid will be feebly dyed in comparison with the other, because the acid has removed part of the aluminium from the tissues (and would remove almost all of it if given sufficient time). The hydrogen ions of the acid compete with the positively-charged hydrated aluminium ions for attachment to negatively charged sites in the tissue. Full details of this test, and of another giving confirmatory evidence, have been published elsewhere.

METHODS OF STAINING

A number of good staining procedures are available. Each procedure has its attributes and many are designed for a specific purpose. Space permits a listing of but a few of the available methods (27-29, 32-35). It is assumed that the reader is familiar with the general techniques for embedding, sectioning and mounting (27-29). It should be recognized that the techniques provided here may require modification in time of fixation, staining and destaining (differentiating) depending on the tissue source, size of tissue and procedure.

A. Wright Stain

This stain is suggested for cells grown in suspension culture. This procedure is a rapid method for evaluating mitotic index and general morphology of the isolated cell.

Materials

1. Wright stain"	0. 1 gm	Dissolve in 60 ml of absolute methanol with the aid of a mortar and pestle.

Procedure

1. Centrifuge a 1-5 ml sample of a cell suspension and decant supernatant fluid.
2. Resuspend in a small drop of BSS and make a film of cells on a slide. Spread thinly as for a blood film and allow to dry in air.
3. Cover preparation completely with Wright stain for 1 minute.
4. Add to the Wright stain an equal quantity of distilled water and counterstain for 2-3 minutes.
5. Wash in running water until clear of any excess stain.
6. A permanent preparation can be made by dehydrating slide through a series of ethanols (50%, 80%, 95%, 100%) and clearing in xylene. Then mount by placing two drops of balsam or HSR on the slide and cover with a coverglass.

B. May-Grunwald-Giemsa Stain

This stain is recommended for monolayer cultures to demonstrate differentially ribo- and deoxyribo-nucleoproteins (RNA-Protein and DNA-Protein). DNA-Protein stains red-purple while RNA-Protein stains blue. Ribonuclease treatment for one hour may be used as a control. MATERIALS

Stock May-Grunwald stain	2.5 gm	Dissolve in absolute methanol to 1000 ml. Age 1 month. Filter.
Stock Giemsa	1. 0 gm	Dissolve in 66 ml of glycerol at 55-60^0 C for 1. 5-2. 0 hrs. ; add 66 ml of absolute methanol.

Procedure

1. Wash in three rinses of warm BSS.
2. Fix for 5 minutes in absolute methanol. Agitate during fixation.
3. Stain for 10 minutes in filtered stock May-Grunwald solution.
4. Stain for 20 minutes in dilute Giemsa solution (dilute 1:15 in distilled water just before use.)
5. Rinse rapidly in distilled water. (10-20 sec.)
6. Quickly rinse in 2 changes of acetone to dehydrate tissue. Do not let coverslip or slide dam.
7. Clear by rinsing three times in acetone-xylol (2:1), three times in acetone-xylol (1:2) and 10 minutes in fresh xylol.
8. Mount in balsam or HSR.

C. Methyl Green-Pyronin Materials

1. Methyl green-Pyronin stain

Methyl green	0. 5%	Dissolve 1 gram methyl green
Pyronin	0. 1 %	in 200 ml of Walpole buffer. Wash dye in separatory funnel with chloroform until all violet color has been removed from stain. Let stand two days until chloroform has evaporated. Add Pyronin to a final concentration of 0. 1%. Store in refrigerator and bring to room temperature before use.

2. Walpole buffer

0. 2 M Glacial acetic acid (HAG)	Add 1. 15 ml Glacial acetic acid to 98. 85 ml water.
0. 2 M Sodium acetate (NaAC)	Add 2. 72 gm Sodium acetate to 100 ml of water.
	Mix 150 ml of 0. 2 M HAC and

50 ml of 0. 2 M NaAC. pH=4.
16.

Procedure

1. Fix in cold FAA for 10 minutes (2 hrs. for paraffin embedded tissues).
2. Rinse in 70% and then 50% ethanol.
3. Rinse in distilled water.
4. Stain for 5 minutes in methyl green-pyronin stain.
5. Rinse in distilled water.
6. Dehydrate with tertiary butyl alcohol-ethanol (3:1) 3X.
7. Clear in two changes of xylene (5 minutes each).
8. Mount in balsam or HSR.

D. Methylene Blue

This procedure is designed for demonstration of nucleoproteins and mitotic figures.

Materials

1. Methylene blue stain

Methylene blue 1% aqueous solution	10 ml
Citric acid - phosphate buffer	
pH 5. 6 (84 ml of M/ 10 citric	
acid; 116 ml of M/5 disodium	
phosphate (anhydrous)	10 ml
Acetone 25 ml Distilled H_2O.	140 ml

2. Rosin 2% solution in 95% ethanol

Procedure

1. Fix in FAA for approximately one hour.
2. Rinse in 50% ethanol.
3. Rinse in distilled H_2O.
4. Stain for 15 minutes.
5. Destain in 95% ethanol plus 2% rosin.

6. Dehydrate through absolute ethanol and xylene.
7. Mount in balsam.

Control slides should be incubated for one hour in 0.01% ribonuclease at 60° C.

E. Hematoxylin and Eosin

This procedure is used routinely for explants which have been embedded and sectioned and is recommended for general histology. Nuclei stain blue, while cytoplasm stains pink.

Materials

1. Alum Hematoxylin

Hematoxylin	0. 5 gm	Dissolve solids in 70 ml distilled water. Add glycerol and acetic acid; if residue remains, filter; remains filter; solution is then ready for use
Ammonium alum:		
$[(NH_4)_2\ SO_4 \cdot Al_2(SO_4)_3 \cdot$	5.0 gm	
$24H_2O]$ $NaIO_3$	0.1 gm	
Glycerol	30.0 ml	
Acetic acid (glacial)	2.0 ml	

2. Bicarbonate

$NaHCO_3$	1.0 gm	Dissolve in 100 ml distilled H_2O.

3. Eosin Y

Dry powder	0. 5 gm	Dissolve in 100 ml H_2O.

Procedure

1. Rinse three times in warm BSS.
2. Fix in neutral buffered formalin for 30 minutes. (If explants, paraffin embed, section, mount and hydrate).
3. Rinse once in distilled water.
4. Stain for 10 minutes in alum -hematoxylin diluted 1:20 in distilled water.
5. Rinse thoroughly in tap water.
6. Expose until cells assume a blue color in 1% aqueous NaHCO

7. Stain in 0. 5% aqueous eosin for one minute.
8. Rinse once in distilled water.
9. Dehydrate rapidly through two changes of acetone to prevent drying.
10. Rinse 3 times in acetone:xylol 2:1.
11. Rinse 3 times in acetone:xylol 1:2.
12. Rinse 3 times in xylol.
13. Clear in xylol (fresh) for 10 minutes.
14. Mount with balsam.

F. Regaud Iron Hematoxylin-Mason Trichrome Method

This procedure is recommended to stain sectioned tissue such as explants or organ cultures. This combination of dyes was designed as a differential stain for collagenous connective tissue and for reticulum. It is also a good stain for general morphology. Fibrous connective tissue and hyaline cartilage are deep blue; muscle, myelin and red blood cells are various shapes of yellow; cytoplasm pink with deep purple nuclei; fibrin is deep pink and bone matrix is bright red.

Materials

1. Picric acid-alcohol solution—95% ethanol saturated with picric acid
2. Masson A—0. 3 gm acid fuchsin
   1 ml glacial acetic acid
   0. 7 gm ponceau de xylidine
   100 ml distilled water
3. Masson B—1.0 gm phosphomolybdic acid
   100 ml distilled water
4. Masson C—2.0 ml glacial acetic acid Glycerol
   100 ml distilled water
   2-3 gm aniline blue or

   Light Green—2 gm Light Green
   S. F. Yellowish

100 ml 1% acetic acid

5. Regaud Hematoxylin—1. 0 gm hematoxylin
   10 ml absolute ethanol
   10 ml glycerine
   80 ml distilled water
   Allow to age three weeks
6. Eosin-Alcohol — Eosin Y in 80% ethanol — 1% solution
7. Iron Alum, (ferric ammonium sulfate), 5% solution

Procedure

1. Fix in FAA for 24 hours.
2. Dehydrate by replacing with 80% ethanol for minimum of 4 hours (may be stored here).
3. Add 5 drops eosin-alcohol to 80% ethanol and allow to stand for ½ hour (to stain small explants so they may be seen in the paraffin blocks).
4. Dehydrate in 95% ethanol for 12 hours.
5. Dehydrate in absolute ethanol for 12 hours.
6. Store in cedarwood oil overnight.
7. Rinse in xylene (or benzene) 2 changes, 15 minutes each.
8. Embed at 60^0 C in a vacuum oven in at least 2 changes of paraffin, hour each or 1 hour each in an oven at atmospheric pressure.
9. Block in paraffin.
10. Section and mount on slides.
11. Place in xylol for 3 minutes.
12. Rehydrate by consecutive changes from absolute ethanol to distilled water, 3 minutes each change (100%, 95%, 70%, 50%, $H2^0$)
13. Mordant in 5% iron alum for 4 hours at room temperature or 5 minutes at 50° C.
14. Rinse in distilled water.
15. Stain in Regaud hematoxylin for 2-3 minutes at 45-50° C.

16. Rinse in 95% ethanol.
17. Destain in picric acid-alcohol and check with microscope. (Nuclear detail should be visible.)
18. Wash in running water for 10 minutes.
19. Stain in Masson A for 5 minutes.
20. Rinse in distilled water.
21. Destain in Masson B for 3-15 minutes.
22. Stain in Masson C for 2-5 minutes or Light Green for 3 minutes.
23. Rinse in distilled water.
24. Destain in Masson B for 5 minutes.
25. Destain in 1% acetic acid for 2-5 minutes.
26. Dehydrate in 2 changes each of 95% and absolute ethanol.
27. Clear in xylene and mount.

CYTOCHEMICAL PROCEDURES

A. Periodic Acid-Schiff (36) and Methylene Blue

The PAS technique is recommended to demonstrate neutral mucopolysaccharides, glycoprotein and mucoproteins and for glycogen when coupled with amylase digestion. The counterstain (methylene blue) demonstrates cellular basophilia (RNA, DNA, acid mucoploysaccharides). It is designed for sectioned material and is adaptable for monolayers as well.

Materials

1. Periodic acid solution

Periodic acid	1. 2 gm	Dissolve periodic acid in 30 ml distilled wat-
Sodium acetate M/5	15 ml	-er, add the sodium
Absolute ethanol	105 ml	acetate and then the
Distilled water	30 ml	ethanol.

2. Reducing rinse

Distilled H2O	60 ml	Dissolve solids in 60

Potassium iodide	3 gm	ml distilled water, stir in the ethanol and
Sodium thiosulfate ($Na_2S_2O_3 \cdot 5H_2O$)	3 gm	then add the HCl.
Absolute ethanol	90 ml	
HCl, 2 N	1. 5 ml	

3. Sulfite wash water

10% sodium metabisulfite ($Na_2S_2O_5$)	30 ml	Add the metabisulfite to 540 ml distilled water and then add
1 N HCl	30 ml	the HCl.
Distilled water	540 ml	

4. Schiff Reagent

Basic fuchsin	2.0 gm	Dissolve stain and metabisulfite in acid.
Potassium metabisulfite ($K_2S_2O_5$)	3.8 gm	Shake at intervals for 2 hrs.
0.15 N HCl	200 ml	Solution should be straw colored. Add
Activated charcoal	500 mg	500 mg fresh activated animal charcoal (to decolorize) and filter. Store in small, dark glass bottles in refrigerator.

5. Methylene blue stain.
6. Amylase solution, 1% in distilled water.
7. Rosin, 2% in 95% ethanol.

Procedure

1. Fix in FAA (or Carnoy)
2. Paraffin embed and section. (If monolayer, skip to step 4).
3. Mount and hydrate (xylene, 100%, 95% ethanol).

4. Three rinses of 70% ethanol.
5. Place in periodic acid solution for 10 minutes.
6. Rinse thoroughly in 70% ethanol (Do not let stand.)
7. Place in reducing rinse for 5 minutes exactly.
8. Rinse thoroughly in 70% ethanol (Do not let stand.)
9. Treat with Schiff reagent for 45 minutes.
10. Rinse in three changes of sulfite wash water, 2 minutes each, exactly.
11. Wash under running water 10 minutes, rinse in distilled water.
12. Stain in methylene blue solution for 2 minutes.
13. Rinse in distilled water.
14. Destain in 2% rosin solution. (Tissue should be pale blue and the blue stain should not come out in clouds.)
15. Dehydrate through two changes absolute ethanol.
16. Pass through four changes in xylol.
17. Mount in balsam or HSR.

To remove glycogen, hydrate through 70% ethanol to water after step 3 and place in unbuffered 1% amylase solution for 15 minutes at 37°C. Rinse thoroughly in distilled water, then in 50% ethanol for 5 minutes and continue with step 4.

B. Periodic Acid-Schiff and Alcian Blue

Neutral and acidic mucopolysaccharides may be simultaneously demonstrated with PAS and alcian Blue. The latter dye stains complex carbohydrates rich in free acidic groups.

Materials

1. Alcian Blue solution

Alcian Blue 8 GX	0.1 gm	Add acetic acid to water and disGlacial acetic acid 3. 0 ml solve dye in, the weak acetic acid Distilled

water 97. 0 ml solution. Filter and add crystal of thymol to preserve.

2. Periodic acid solution.
3. Reducing rinse.
4. Sulfite wash water.
5. Schiff Reagent.
6. Alum Hematoxylin.
7. Neutral buffered formalin.
8. 3% acetic acid.

Procedure

1. Fix in neutral buffered formalin.
2. Paraffin embed and section. (if monolayer, skip to step 4)
3. Mount and hydrate (xylene, 100%, 95%, 70%, 50% ethanol, water).
4. Rinse in 3% acetic acid for 3 minutes.
5. Stain 2 hours in Alcian Blue solution.
6. Rinse in tap water and then in 3% acetic acid for 3-5 minutes.
7. Wash for 2 minutes in running tap water and then briefly in distill water.
8. Follow steps 5-11 under Periodic Acid Schiff and Methylene blue.
9. Stain 5 minutes in Alum Hematoxylin.
10. Wash 2 minutes in running tap water.
11. Dehydrate through 70%, 95% and 2 changes absolute ethanol followed by 2 changes of xylol.
12. Mount in HSR or balsam.

C. Feulgen Technique for DNA

This procedure is perhaps the most specific staining technique available for demonstrating deoxyribonucleoprotein.

Materials

1. HCl, 1 N
2. Schiff reagent.
3. Sulfite wash.
4. Fast Green FCF, 0. 01% in 95% ethanol

Procedure

1. Fix in Carnoy fixative for 20-30 minutes.
2. Rinse in absolute ethanol for 3-5 minutes.
3. Hydrate through 95% and 70% ethanol.
4. Rinse in distilled water.
5. Incubate in 1 N HCl at 60°C for 10 minutes.
6. Rinse in distilled water.
7. Stain in Schiff reagent for 10 minutes. .
8. Rinse in three changes of sulfite wash, two minutes each.
9. Wash in running water for 15 minutes or until sulfite smell is gone.
10. Rinse in 70% ethanol.
11. Counter stain in Fast Green FCF solution for 5 seconds.
12. Dehydrate in absolute ethanol and clear in xylene.
13. Mount in balsam or HSR.

D. Propylene Glycol-Sudan Black Technique

This procedure is recommended for the demonstration of mitochondria, phospholipids as well as neutral fats in monolayers. Lipids are stained blue-black.

Materials

1. Sudan Black Solution

Sudan Black B	0. 7 gm	Dissolve Sudan Black in propylene glycol
Absolute propylene glycol	100.0 ml	at 100-110°C. Do not exceed this temperature. Filter hot

		through Whatmann No. 2 filter paper. Cool and refilter with vacuum through medium porosity fritted glass filter.

2. 85% propylene glycol
3. Polyvinylpyrrolidone (PVP) Mounting Medium (38)

   (Refractive index 1.43)

Polyvinylpyrrolidone	50 gm	Dissolve PVP in distilled
Distilled water	50 ml	water, let stand over
Glycerol Thymol, crystalline	2 gm	night. Add glycerol and stir mixture thoroughly. A crystal of thymol can be added as a preserva- -tive.

Procedure

1. Fix in cold neutral buffered formalin for 10 minutes.
2. Rinse in water for 2-5 minutes.
3. Rinse in propylene glycol for 3-5 minutes.
4. Stain in Sudan Black B for 5-7 minutes.
5. Destain in 85% propylene glycol for about 2-3 minutes.
6. Wash in distilled water for 3-5 minutes.
7. Mount with polyvinylpyrrolidone.

NOTE: Cold acetone extraction for 30 minutes prior to this procedure will remove all lipids other than phospholipid.

E. Baker Acid Hematin for Phospholipid Materials

1. Formol-Calcium fixative

Formalin (40% Formaldehyde)	10 ml	Mix ingredients in
10% aqueous $CaCl_2$ (anhydrous)	10 ml	distilled water. Add
Distilled water,	80 ml	excess of $MgCO_3$ to neutralize.

2. Dichromate-Calcium solution

Potassium dichromate	5 gm	Dissolve ingredients in distilled water. Solution is
Calcium chloride (anhydrous)	1 gm	stable. A small
Distilled water	100 ml	precipitate may be ignored.

3. Baker acid hematin

Hematoxylin	50 mg	Dissolve the hematoxylin
Potassium or Sodium iodate		in distilled water. Add the iodate and
(1% solution)	1 ml	heat just to a boil.
Glacial acetic acid	1 ml	Cool and then
Distilled water	48 ml	add glacial acetic. Prepare on same day of use.

4. Borax-ferricyanide solution

Potassium ferricyanide	0.25 gm
Borax ($Na_2B_40_7 \cdot 10\ H_2O$)	0.25 gm
Distilled water	100 ml

5. Weak Bouin fixative

Saturated aqueous picric acid	50 ml	Add the glacial acetic acid just before use.
Formalin (40% formaldehyde)	10 ml	
Distilled water	35 ml	
Glacial acetic acid	5 ml	

6. Pyridine, USP

Procedure

1. Rinse cover slips in BSS.

2. Fix in formol-calcium for 18 hours.
3. Incubate in dichromate calcium solution for 18 hours at 28° C.
4. Incubate in dichromate calcium solution for 24 hours at 60° C.
5. Wash in distilled water for 5 minutes, changing water repeatedly.
6. Stain in freshly prepared Bakers acid hematin for five hours at 37° C.
7. Rinse in distilled water.
8. Differentiate in a Borax-ferricyanide solution for 18 hours at 37° C.
9. Rinse repeatedly in distilled water.
10. Mount in Polyvinylpyrrolidone.

Baker Acid Hematin Control

1. Rinse coverslips in BSS.
2. Fix in weak Bouin for 20 hours.
3. Wash coverslips in water (3 changes, 5 minutes each).
4. Place in pyridine solution for 1 hour at 28° C.
5. Place in fresh pyridine solution for 24 hours at 60° C.
6. Wash cover slips in dichromate calcium solution for 18 hours at 28° C.
7. Proceed as in Baker Acid Hematin Test for phospholipids (Step #5).

F. Alkaline Phosphatase Technique

This histochemical procedure is included as one example of many such techniques that are becoming increasingly available. The procedure recommended here is the azo-dye method in view of its simplicity. Sites of enzyme activity appear as dark blue granules.

Materials

1. Reaction mixture

Naphthyl ASE phosphate	20 mg	Dissolve the naphthyl ASE phosphate
N, N-dimethyl formamide	0. 5 ml	in dimethyl formamide.
Fast Blue RR	60 mg	Add 50 ml of the bu-
$MgSO_4$ anhydrous	60 mg	ffer. Add $MgSO_4$
0. 1M Tris Buffer (pH 8. 8)	100 ml	Add rest of buffer. Then Add Blue RR and shake thoroughly.

2. PVP mounting medium.

Procedure

1. Fix for 5 minutes in cold neutral buffered formalin.
2. Rinse in distilled water 1-2 minutes.
3. Incubate in freshly prepared reaction mixture for 20 minutes.
4. Rinse in distilled water.
5. Mount coverslip on slide with PVP.

NOTE: As a control, use the diazonium salt for 20 minutes without substrate.

G. Non-Specific Esterase Technique

This is another histochemical procedure for a group of enzymes that is being increasingly studied. Sites of enzymatic localization stain red.

Materials

1. Reaction mixture

Absolute propylene glycol	20 ml	Dissolve dye
Sorensen phosphate buffer pH 7.6	20 ml	(Garnet G. B. C.)
12 ml of M/15 $NaH_2PO_4 . H_2O$ and		in distilled water.
88 ml of M/15 Na_2HPO_4 (anhydrous)		Dissolve substrate
Distilled water	58 ml	in buffer. Mix and
1% naphthyl AS-Acetate in 6%	2 ml	acetone add remin-

(substrate)	-ing ingredients.
Garnet G. B. C. salt	50 mg

2. Polyvinylpyrrolidone (PVP).

NOTE: As a control incubate without substrate.

Procedure

1. Fix in cold neutral buffered formalin for 5 minutes.
2. Rinse in distilled water for 1-2 minutes
3. Incubate in substrate for 15 minutes.
4. Rinse in distilled water.
5. Mount in PVP.

H. Lactic Dehydrogenase

The demonstration of lactic dehydrogenase (LDH) activity in cultivated cells is provided as a representative example of the general procedure for demonstrating pyridine nucleotide linked oxidative enzymes.

Materials

1. Monolayer of cells on coverslip (9 × 22 mm).
2. Ten ml Hanks BSS.
3. Two columbia staining jars.
4. Lactate stock solution

Sodium (DL) lactate	5. 0 gm	Dissolve in 5 ml distilled water. Adjust pH to 7. 4 with N NaOH. Bring to final volume of 10 ml. Store in refrigerator.

5. Cyanide solution

Sodium cyanide	4. 0 gm	Dissolve in 5 ml distilled water. Adjust pH with N HCl to pH 7. 2. MUST BE DONE IN VENTILATED HOOD! HIGHLY POISONOUS GAS EVOLVES. Bring to 10

		ml final volume. Store in refrigerator.
6. PMS solution		
Phenazine methosulfate	80 mg	Dissolve in 100 ml distilled water. Store in dark bottle in refrigerator.
7. Reaction mixture		
Diphosphopyridine		Just before use, dissolve DPN and Nitro BT in
nucleotide (DPN)	2.5 mg	Phosphate Nitro B. T.
	4. 0 mg	buffer. Add 0.1 ml of lactate stock solution and 0.1 ml of cyanide solution; immediately Sorenson's phosphate before use add 0.1 ml of PMS solution
Buffer (0.1 M, pH 7.4)	8 ml	

8. Ten ml neutral buffered formalin.
9. PVP Mounting medium.

Procedure

1. Remove coverslips from culture tubes, rinse in warm BSS.
2. Incubate in freshly prepared Reaction mixture for 5- 15 minutes.
3. Rinse briefly in distilled water.
4. Fix in 10% neutral buffered formalin for 10 minutes.
5. Rinse in distilled water.
6. Mount on slides with PVP.

ENUMERATION OF CHRMOMOSOMES

Analysis of the chromosomal complement of cells in culture is becoming an increasingly important tool in genetic studies as well as for the characterization of cell lines. Characterization on the basis of chromosome number alone is not reliable generally since most cell lines become heteroploid in culture. In several instances morphologically distinctive chromosomes have emerged in a cell population which serve to characterize the cell line. The distribution of metacentrics, telocentrics, minutes, etc. also serves to characterize cell populations.

A certain amount of experience with the procedures of chromosomal analysis is required before reliable results can be obtained. It is wise to check your technique with material which has been characterized adequately.

Materials

1. Monolayer culture of L-M strain mouse cells in 199 peptone. These cells should be in early or mid log phase of growth.
2. Sterile colchicine (0. 1 mg/ml in BSS). Sterilize by filtration and store at 4°C.
3. Sterile hypotonic solution (1% sodium citrate in distilled water). Prepare fresh and use warm at 37° C.
4. One 15 ml conical centrifuge tube.
5. One sterile 1 ml cotton plugged serological pipette.
6. Two 10 ml and four 5 ml serological pipettes.
7. One rubber policeman.
8. Acetic acid-methanol fixative (1 part glacial acetic acid and 3 parts absolute methanol).
9. Seventy per cent ethanol.
10. Acetic Orcein stain (Dissolve 0. 5 grams of Gurr natural orcein in 100 ml of boiling 45% acetic acid, cool and filter 3 times). Filter before use if crystals have formed.
11. Microscope slides (1 × 3") and 22 mm square coverslips.
12. Bibulous paper.
13. Small metal spatula.
14. Kronig cement".

Procedure

1. With a sterile 1 ml serological pipette add 0. 1 ml of colchicine solution to a monolayer culture of L-M cells in early log growth phase. Incubate at 35°C for eight hours
2. After eight hours incubation, harvest the monolayer by scraping with a rubber policeman.
3. With a 10 ml serological pipette triturate 3-4 times to disperse the cells and transfer the cell suspension to a 15 ml conical centrifuge tube.
4. Centrifuge the cells gently (150G) for 5 minutes and pipette off the medium.
5. Add 3 ml of warm sodium citrate solution and resuspend. Let stand for 6-8 minutes at 37° C, then add 1 drop of acetic-methanol fixative.
6. Centrifuge for 5 minutes at 150G and pipette off the supernatant fluid. Resuspend the cells in 1 ml of acetic acid-methanol fixative for each ml of the original suspension and let stand at room temperature for 10 minutes.
7. Centrifuge and change the fixative two more times. Finally resuspend in fixative.
8. Dip a clean glass slide in 70% ethanol and immediately drop one or two drops of cells suspended in fixative on the wet slide and pass it quickly through a flame to ignite. (This will spread the chromosomes). Blow to complete drying. Allow the slide to stand for 10-15 minutes before proceeding to the next step.
9. Add one or two drops of acetic-orcein stain to the slide and gently lower a coverslip onto the stain taking care not to introduce bubbles. Gently blot between bibulous paper and seal the edges with Kronig cement, applied with a heated metal spatula.
10. Examine the stained preparation with an oil immersion phase contrast objective. Select a chromosome spread which is well separated from other cells but with the cell membrane intact to insure that you are dealing with a sin-

gle cell. Count the number of chromosomes. Repeat for at least 10 additional cells. (For reliable statistics it is necessary to examine 50-100 spreads).

11. With a camera lucida trace the outlines of several of the chromosomes in a well spread preparation.

PHASE AND INTERFERENCE MICROSCOPY

Observation of specimens with the light microscope is dependent upon the relative absorption of light, a function of the thickness of the specimen and the refractive indices of the specimen and mounting medium (optical path). As a consequence, cells or tissues examined with the light microscope are best visualized when they have been fixed and stained to emphasize specimen contrast. Bright field, dark field or polarized light (for doubly refractive materials) have been employed for the study of living cells where specimen details consist of small differences in optical path. Although these methods provide contrast between the preparation and the mounting medium, they fail to produce contrast within the specimen. Vital staining techniques also may be used but this procedure is time consuming and may cause severe distortion. The phase-contrast microscope and the interference microscope permit visualization of living cells.

The phase-contrast instrument is equipped with phase retardation plates which emphasize the refractive differences between microscopic objects and the medium in which they are immersed. Light rays traverse equal distances through air, water, and glass at different rates and emerge out of phase. These phase differences are not detectable by the eye which responds primarily to intensity differences. The phase-contrast microscope converts these phase differences into intensity differences thus rendering the various indices of refraction visible. Phase differences are enhanced by the choice of a mounting medium with a low index of refraction, such as saline solutions, and the instrument is therefore particularly adapted for studying living tissues.

Interference microscopy can be accomplished by the recombination in the image of two beams of light from the same source, one having been modified by passing through the specimen and the other through the surrounding medium. This microscope provides variable color contrast with white light illumination and intensity variations with monochromatic light. Thus, interference microscopy can be used to measure the thickness of the optical path and mass of the specimen. For purely observational purposes, the phasecontrast microscope is recommended whereas interference microscope is essential for accurate measurement of phase changes which permits measurement of mass, optical path, etc.

Cells from monolayers or suspension cultures may be examined satisfactorily in normal balanced salt solution with the phase or the interference microscope. The thickness of specimen which can be examined depends on its transparency, and its optical density. Dense tissues or materials up to about 4 microns thick may be used and less dense specimens with a thickness of 100 microns or more can be studied. Large specimens such as explants must be embedded and sectioned with a microtome. Coverglasses about 0. 18 of a mm thick are essential for best results as objectives in most

phase and interference microscopes are corrected to this standard. A box of #12 coverslips will contain more 0.18 mm coverglasses than will a box of #l or #2 coverslips. It should be recognized also that hollow ground slides (unless ground with a flat surface) and hanging drop mounts will act as lens systems and therefore are not suitable for phase or interference microscopy. Special quartz slides may be used. Similarly, the split-beam interference microscope reveals the total optical path which includes the slide, the mounting medium, the specimen and the coverglass. Any unevenness, defects in the glass, or inhomogeneity in the mounting medium will produce artefacts in the preparation. Some distortion (wedge effect) is usually present in the preparation even with better than average quality coverglasses and slides and when care is used in mounting the specimen. The distortion in interference microscopy will manifest itself as a change of luminescence as the stage is turned or in a different reading depending on the orientation of the sample.

PREPARATION OF SERUM AND PLASMA

Several generalizations concerning the use of serum for growth of cells can be made. Homologous serum provides best growth of cells or organs in culture. If heterologous serum must be used, cytotoxicity increases in proportion to the concentration of serum employed. Freshly drawn serum is superior to serum stored for weeks at 4 or –20° C. It is most economical to use commercial serum for routine cell culture techniques. Lyopholized sera, because of the loss of CO_2, is alkaline and often requires gassing with CO_2 to bring it to a physiological pH, viz., 7.2 or 7.4. Because of cytotoxicity or nonspecific virus neutralization, contributed by the "normal" antibody-complement system, it may be prudent to heat-inactivate sera (56° C for 30 minutes) prior to use. Some of the variations in the growth qualities of sera are known to be attributable to differences among individual donors or/and to seasonal or dietary factors which are difficult to control. Because of these contingencies sera may be pooled to minimize such variations. Recent knowledge concerning the selective action of sera on variants in cell populations suggests caution in choice of serum where population changes may be of importance.

Plasma clot procedures still are used in a number of laboratories for establishing explant cultures or for deriving established cell strains. Since many commercial preparations of heparin contain phenol or some other toxic preservative, heparin from Connaught Medical Research Laboratories (Toronto, Canada) is recommended.

Material (human venipuncture)

NOTE: State laws may require supervision of bleeding of human subjects by an attendant physician.

1. One sterile 30 ml hypodermic syringe.
2. Two sterile 19 gauge, 2 inch, short bevel hypodermic needles.
3. Cotton plegets moistened with 70% ethanol for skin disinfection.
4. Cotton plegets moistened with tincture of iodine (20%) for skin disinfection.

5. One flexible rubber hose to be used as a tourniquet.
6. One sterile, screw-cap, 40 ml, conical centrifuge tube.
7. One sterile gauze pad to be used as a compress.
8. Media for sterility checks.
9. One "propipette".
10. Three sterile 16 × 125 mm screw-cap tubes. 11. Sterile applicator sticks.

Materials (to obtain cockeral blood)

1. One sterile, 30 ml syringe (oiled before autoclaving with USP mineral oil, 4% in petroleum ether; alternatively, syringes may be rinsed with heparin (0.2 mg%) prior to use to prevent clotting).
2. Two sterile 20 gauge, 3 inch, short bevel hypodermic needles.
3. Plegets as described above.
4. One sterile, screw-cap, 40 ml conical centrifuge tube containing 0.3 mg of heparin.
5. One sterile gauze pad to be used as a compress.
6. One "propipette".
7. Media for sterility checks.
8. Sterile applicator sticks.
9. Three, sterile, 16 × 125 mm screw-cap tubes.

Procedure

1. Carefully disinfect the skin and draw 30 ml of human blood by venipuncture (median cubital vein) as demonstrated by the instructor. Aseptically transfer the blood to the centrifuge tube, permit it to clot, and free the clot from the walls of the tube with a sterile applicator stick.
2. Bleed the rooster from the wing vein or by cardiac puncture as demonstrated. (The head of the bird should be covered to minimize restlessness; a sharp needle and a good light source are needed to facilitate entry into the vein;

blind probing is worse than useless.) Mix the blood and heparin thoroughly to prevent clotting. Press the gauze pad over the site of puncture and hold it firmly in place for at least five minutes; avian blood clots slowly.

3. Centrifuge all specimens for 30 minutes at approximately 2500 RPM; aspirate supernatant fluids aseptically by use of a 10 ml pipette and a rubber bulb. A "propipette" bulb permits ready removal of supernatant fluids without resuspending cells.
4. Dispense the serum and plasma in 10 ml aliquots into the 16 × 125 mm tubes.
5. Inoculate culture media for sterility checks.
6. Store the serum and plasma at 4° C.

PREPARATION OF EMBRYO EXTRACT

Although established cell strains propagate readily in balanced salt solutions to which serum has been added, it may be expedient if not essential to supplement culture media with growth stimulatory substances for isolation of cell strains or to establish explant or organ cultures. Extracts of tissues from a wide range of animal sources have been used for this purpose. Because embryonic tissue usually is rich in growth stimulatory substances, it is used in preference to extracts of adult tissues. If tissue differentiation is germane to experimental studies, age of the embryos used for preparation of extracts must be considered, e.g., extracts from six to nine day old chicken embryos significantly promote growth while those from twelve to fourteen day old embryos influence tissue differentiation and growth. Chicken embryos from nine to fourteen days of age are the most convenient source material for preparation of embryo extract. When large quantities of material are required, or if mammalian embryos are employed, tissues can be minced in a Waring type blendor. Because embryo extracts can serve as a source of somatic cells that may contaminate cell strains, it may be necessary to filter such additives as a precautionary measure.

Embryo extract may serve as a source of bacterial contaminants, viz., Salmonellae, or may contain viruses or other filterable microbes, e.g., PPLO which are not readily detectable. Effective use of embryo extracts in tissue culture research requires an appreciation of these problems as they relate to interpretation of experimental findings.

Materials

1. Three, nine to twelve day old embryonated chicken eggs.
2. Waxed egg carton to hold eggs upright.
3. Cotton plegets moistened with 2% tincture of iodine.
4. Cotton plegets moistened with 70% ethanol.
5. Two fine-tip forceps.
6. Two sterile Petri dishes.
7. One sterile, 30 ml syringe, the body equipped with a 28 gauge stainless steel wire mesh; the body and plunger of the syringe should be wrapped separately and autoclaved.
8. One sterile screw-cap, 40 ml conical centrifuge tube.
9. Three, sterile, screw-cap, 16 × 125 mm tubes.
10. Egg candler.
11. Fifty ml BSS.
12. Culture media for sterility checks.

Procedure

1. Candle the eggs and select those which are fertile.
2. Place the fertile eggs, air-sac end uppermost, in the waxed egg cartons.
3. Disinfect the egg shells with the iodine and ethanol plegets as demonstrated by the instructor. Gently crack the shell over the air-sac; remove the shell surrounding the air-sac with sterile forceps.
4. Tear off the shell membrane, gently grasp the neck of the embryo and exert gentle upward traction until the embryo is free of attached membranes. (If forceps are compressed

too tightly or if excessive traction is applied, the neck will break and the embryo will be lost.)

5. Transfer the embryos to a Petri dish or directly to the 30 ml syringe.
6. Insert the plunger into the syringe, measure the compressed volume of the pooled embryos, and then forcibly extrude the embryos through the screen.
7. Collect the minced embryonic tissue in the conical centrifuge tube and aseptically add an equivalent volume of BSS.
8. Mix the minced tissue and BSS thoroughly; centrifuge the embryo extract for 30 minutes at approximately 2500 RPM.
9. Dispense the supernatant fluid into the 16 × 125 mm tubes and inoculate culture media for sterility checks. Store the embryo extract at –20° C.

NOTE:

1. Thawed embryo extract usually must be centrifuged before use and the sedimentable material discarded.

2. Fresh, frozen or lyophilized embryo extract can be used to coagulate plasma; ultrafiltrates of embryo extracts are not satisfactory for this purpose.

PRESERVATION OF CELLS

For many years the only method for maintaining tissue cell lines was by serial subculture at frequent intervals. By this method Carrel maintained his celebrated culture of chick heart fibroblasts for over 34 years. Several groups of workers have demonstrated the feasibility of long term storage of animal cells at low temperatures. This has greatly simplified the problem of cell preservation. A variety of primary cells, cell strains and established cell lines have been shown to survive when stored at –65° C or below without a discernible change of properties.

While satisfactory survival has been attained with storage in the range of –65° C (electric deep freeze) to –79° C (solid carbon dioxide chest), it is recommended that the storage temperature not be above –90° C. Storage in liquid nitrogen (–196° C) or in the vapour above liquid nitrogen (–150° C to –180°C) are the most practicable methods currently available for achieving such temperatures.

There are relatively few hazards involved in working with liquid nitrogen. Direct contact with the skin is to be avoided as it may result in frostbite. Work with nitrogen should always be in a well ventilated room to avoid anoxia though the danger is not great under ordinary circumstances. While liquid nitrogen is not explosive per se, a danger is involved when containers are immersed in liquid nitrogen. Unless all immersed containers are watertight, liquid nitrogen may leak in and when the containers are removed from storage, vaporization of the nitrogen will lead to rapid expansion with a resulting explosion of the container. This is most likely to occur with ampoules which are improperly sealed o with stoppered or screw-cap containers. Storage in the vapor phase prevents such an occurrence.

Adequate attention must be given to procedures for freezing and thawing and to the menstrum in which the cells are suspended. The critical points in technique are: l) slow freezing, 2) rapid thawing, 3) use of 5-20% (v/v) glycerol or 5-10% (v/v) dim ethyl sulfoxide in the freezing and storage medium, and 4) storage at temperatures below –90° C.

The freezing and storage medium normally is the growth medium for the particular cells being frozen. To this is added 5-20% glycerol or 5-10% dimethyl sulfoxide. The exact amount added will depend both upon the particular cells and upon the amount of serum in the medium. Freezing should take place at approximately 1°C drop per minute down to –30°C. The temperature can then. be dropped rapidly by placing the sample in liquid N_2 or in the vapor above liquid N_2. Thawing Should be done in a water bath at +45°C with agitation and should require 1 ss than 1 minute.

Glycerol or dimethyl sulfoxide should be added to the culture medium before suspending the cells in the medium or should be added gradually to the cell suspension. Upon thawing, the cell suspension should be diluted to the desired concentration of cells by addition of fresh medium. This should be done gradually. The final concentration of glycerol or dimethyl sulfoxide should be less than 1%. Under no circumstances should the cell suspension be centrifuged before plating as this will cause unnecessary trauma.

Several methods are available for attaining either an accurate freezing rate or a satisfactory approximation of a 1°C/minute drop in temperature. The most elaborate method involves the use of an electronic device for programming the temperature drop by controlling the rate at which liquid nitrogen vapor is introduced into a sample storage chamber. A less sophisticated method freezes samples in the vapor contained in the neck of a liquid nitrogen storage tank. By regulating the position of the samples in the neck of the tank, different freezing rates can be achieved. A less accurate method is to place ampoules at 4°C (ordinary refrigerator) for 2-4 hours, then into the –20°C freezing compartment for 4-24 hours and finally into the storage chest.

Ampoules for freezing should be of borosilicate glass with a low coefficient of expansion. Sealing should be done by the pull method, rather than by fusion, using an oxygen flame. An automatic pull sealer is available commercially. Ampoules may be tested for proper sealing by immersing them in strong dye solution or in 95% ethanol. In the latter case a leak will be indicated by precipitation of protein in the cell suspension.

If gummed labels are used these should be autoclaved on the ampoules to prevent loss on storage. If masking tape is used this procedure is not necessary. Ampoules may be filled with a hypodermic syringe using a large bore needle of adequate length (e.g. 15 gauge, 3Z").

A typical protocol for preparation of L-M mouse cells for freezing is presented below. With suitable modifications of medium and harvest procedures the method is applicable to many other cells.

Materials

1. Five 2 oz. French square monolayer cultures of L-M cells in 199 peptone. These cells should be in the mid-logarith-

mic phase of growth.

2. Five sterile rubber policemen.
3. Sterile reagent grade dimethyl sulfoxide (sterilize by autoclaving).
4. Sterile 40 ml screw-cap conical centrifuge tube.
5. Twenty ml sterile 199 peptone.
6. Ten sterile 1 ml ampoules.
7. One sterile 10 ml hypodermic syringe and one sterile 15 gauge, 3Z inch hypodermic needle.
8. Six 5 ml, two 10 ml and two 1 ml sterile cotton-plugged serological pipettes.

Procedure

1. Using sterile rubber policemen scrape the monolayers into the medium in which they are growing. Triturate with a pipette to disperse the cells and transfer the cell suspension to the sterile 40 ml centrifuge tube. Pool the cells from the five bottles.
2. Centrifuge the cell suspension at 500 RPM for 10 minutes and then decant the medium.
3. Add 10% (v/v) of dimethyl sulfoxide (DMS) to 20 ml of 199 peptone. Mix thoroughly.
4. Add 5 ml of the 199 peptone (containing DMS) to the sedimented cells and resuspend them. Count the cells and adjust the concentration to ?X106/ml by diluting with 199 peptone (containing DMS) as required. Do a viable count using the erythrosin B method and record the results.
5. Take up the cell suspension in the sterile 10 ml hypodermic syringe fitted with a 15 gauge, 3z inch hypodermic needle. Dispense 1 ml of cell suspension into each sterile ampoule.
6. Seal the ampoules by pulling out the neck in an oxygen flame.
7. Label each ampoule with the proper strain designation, number of cells/ml, medium and preservative and the date.

Upon recovery of the cells from the freezer, total and viable cell counts should be made and compared with the counts at the time of freezing to determine percentage survival. With cell lines which clone well, more meaningful data can be obtained by determining plating efficiency before and after storage.

STAINING FOR ELECTRON MICROSCOPY

Staining for electron microscopy means increasing specimen contrast by depositing in certain areas atoms of high atomic number which have a greater electron scattering power than the atoms (C, O, N, H) of the tissue. Osmium tetroxide and potassium permanganate stain during fixation by selective deposition of their heavy atoms: Os and Mn. Osmium increases the contrast of phospholipid membranes, proteins and unsaturated lipids; permanganate stains only membranes and lignin, most other components being destroyed and removed.

Staining During Dehydration

Potassium permanganate can be used to increase the contrast of membranes. Stain for 10-15 minutes with 1 $KMnO_4$ in absolute acetone after dehydration. Wash briefly in two changes of acetone with a reducing agent added (2 drops methyl acrylatel to 25 ml acetone). Wash briefly in acetone twice more, then embed.

Phosphotungstic acid (PTA), once the only common stain in electron microscopy, can also be used during dehydration. A 0.1-1% solution in 70% alcohol for 30 minutes will give a general enhancement of contrast, probably by staining proteins.

Uranyl acetate (UA) is much like PTA, but stains nucleic acids in addition to protein. It is used as a 1-2% solution in 75% alcohol. The tissue can be left several hours, even overnight. UA can also be used in the final absolute alcohol (1-2%) and the tissue left immersed for 1 to 2 hours.

Staining Thin Sections

Treatment of the sections themselves allows the same material to be stained in several ways. This permits a comparison of results.

Most stains penetrate best if the sections are not allowed to dry after removal from the microtome bath before staining. Submerge the grids, with the aid of forceps, in drops of the chosen stain placed in the bottom of covered petri dishes or in larger volumes of stain in covered watch glasses. When using lead hydroxide preparations, place a few beads of caustic potash' or similar CO_2 removing agent in the dish along with the drops to reduce CO_2 contamination. Do not breathe over the drops. The citrate preparations are recommended as this greatly reduces risk of CO_2 contamination. Do not allow the grids to dry between staining and washing. This will cause deposition of a contaminating layer. Sections are washed by holding the grid in forceps and flushing with drops of distilled water from a wash bottle.[5] Place them on a filter paper, sections upwards, to dry.

Uranyl acetate. This is used as a saturated solution in 50% alcohol, brought to pH 4-5 with NaOH, or a 2% aqueous solution without adjusting the pH. Stain for 30-90 minutes.

Lead citrate. This enhances the contrast of nucleic acids, membranes and particularly glycogen, which is otherwise difficult to stain. The citrate salt is the most useful form as it reduces CO_2 contamination. Add 1.33 gm $Pb(NO_3)_2$ and 1.76 gm Na_3 $(C_6H_5O_7)$ $2H_2O$ to 30 ml water, shake well for one minute, stand for 30 minutes with occasional shaking. Add 8 ml 1N NaOH, dilute suspension to 50 ml. The stain is ready for use when the suspension dissolves. Stored in polythene bottles it will keep for several months. When staining, place the drops on polythene or wax surfaces to maintain surface tension; 30 minutes should be sufficient.

Lead hydroxide. This is the original lead stain and is used in the same way as lead citrate. Special care must be taken to avoid CO_2 contamination; always keep the dish covered and do not breathe over the drops. Make the stain by adding an excess of concentrated ammonia to 50 ml of hot concentrated lead acetate. Stir the paste in one litre of distilled water, decant and rewash with another litre. Collect the precipitate. Do not allow the paste to dry. Saturated solutions are made by adding a little paste to water or methyl alcohol, shaking and decanting the clear liquid. This stain is very active and staining time is important as extraction of cell constituents may take place. 30 minutes should be sufficient.

Potassium permanganate. This is sometimes useful as a stain for sections. Use drops of a freshly prepared 2% solution in water as the stain, as a film of reduced compounds forms quickly on an exposed surface. Make the drops using a fine pipette and place the grids in them immediately. If a granular deposit forms on the surface of the section, it can be removed with a weak solution of citric acid.

Double Staining

The use of two stains, one after another, to increase contrast uniformly can be valuable. Some combinations are

Uranyl acetate and lead citrate

Uranyl acetate and lead hydroxide

Uranyl acetate and potassium permanganate

Potassium permanganate and lead hydroxide

STAINING OF RESIN-EMBEDDED SPECIMENS FOR THE LIGHT MICROSCOPE

It is often very convenient to be able to look at sections under the light microscope to decide the orientation of a piece of tissue in a block or to locate a particular cell or group of cells in a tissue. Resin-embedded sections cannot be treated in the normal way for light microscopy, but the following technique will produce a satisfactory preparation for the light microscope.

Preparation of AMB Stain

(A) 1 g Methylene Blue and 1 g sodium borate (borax). Dissolve the borax in a little distilled water, add the Methylene Blue and make up to 100 ml with water.

(B) Dissolve 1 g Azur 11 in 100 ml of distilled water. Mix A and B 1: 1 just before use. (A and B do not keep long once mixed.)

Cut sections with green transparent interference colours. Pick them up from the water bath with an eyelash or stiff whisker. Sections that are too thin will wrap themselves around the whisker when they are lifted. Transfer the sections to a drop of distilled water on a microscope slide; several sections to a drop (Figure 11). Allow the drop to evaporate slowly over a low bunsen flame

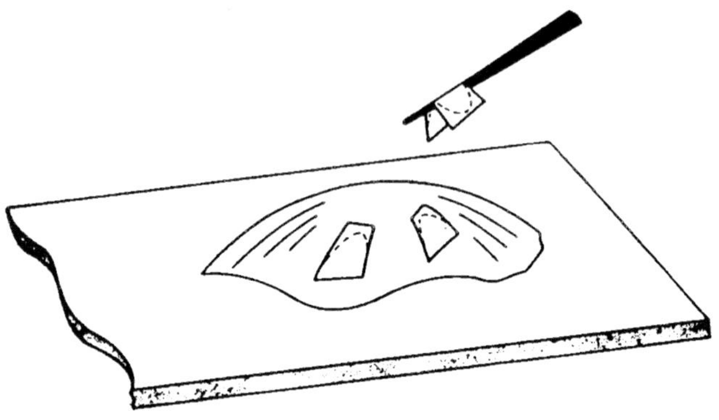

Figure 5.1

or on a hot plate, leaving the sections stuck on the slide. Cover the sections with 1% AMB stain in 1% borax. Warm the slide again as before for 1-2 minutes, but do not dry down the drop of stain. Wash away the excess stain with tap water and finally rinse clean with distilled water.

CYTOCHEMICAL STAINING

Cytochemistry attempts to locate specific chemical sites within the cell. The general stains used in electron microscopy often have some chemical specificity, but do not locate individual sites with any certainty. Electron microscope cytochemistry is still in its infancy, especially as regards plant tissue. Many reactions have been used in animal tissue which are not yet usable in plants.

Some cell components have sufficient electron density to be visible without staining. Nucleic acids are visible in tissues fixed in acetic alcohol, although neither the fixation nor their contrast is really adequate. The coloured matter in many brown and red pigment cells is very electron dense, since it has a high iron content, and appears satisfactorily dark even in stained sections. In all other cases some way of binding a heavy metal to a particular substance, or of making it precipitate a heavy metal, is needed to give the substance electron contrast.

Like general stains, cytochemical stains can be applied either to the wet tissue blocks after fixation, or to the resin-embedded sections. The fixative used must be chosen to suit the particular cytochemical technique. Glutaraldehyde is often the best and is essential if enzyme activity is to be preserved. Dehydration may leach out or otherwise affect some components and resin may impede the penetration of reagents into sections. The development of watersoluble embedding media has greatly reduced the importance of these latter two problems and many techniques, formerly only usable on bulk tissue, may well become feasible for sections. This would be expected to give better preservation of ultrastructure when using strong reagents in the staining technique.

NUCLEIC ACIDS

Bismuth staining

Bismuth forms complexes with organic (or inorganic) phosphate to give a very insoluble and electron dense precipitate. It thereby fulfils the essential requirement of staining nucleic acids but not proteins. In practice, it stains DNA strongly, RNA less so, and also gives an increase in general contrast. No ultrastructural damage is caused. The staining is carried out after fixation, before dehydration. Sections can subsequently be stained with lead. The two stains have complementary effects, and after this double staining the cell wall appears much darker than when either stain is used alone. Chromatin appears both denser and more granular than when only stained with bismuth.

Dissolve 20 mg of metallic bismuth in 0.2 ml of concentrated nitric acid. Take care that the bismuth is covered by the acid. (Bismuth is rapidly oxidized by air in the presence of acid.) Dilute the solution to 80 ml and add 4 ml of 0.1M citric acid to buffer it. Adjust the solution to pH 7 with 1N sodium hydroxide Slight cloudiness may appear at pH 2, but this should disappear again at pH 4-5. Finally make up the solution to 100 *ml* and add sucrose' to a concentration of 0.2M. The stain can be stored in a refrigerator.

After fixation, wash the tissue segments in 0.1N nitric acid buffered to pH 7 with citrate (i.e. the staining solution without bismuth). Then stain them for 90 minutes at 0°C.

Other nucleic acid stains

Iron, as 0.1M $FeCl_3$, has been used, after fixation, as a nucleic acid stain, but it tends to destroy some ultrastructure. Indium can be used, but the technique is complicated and inconvenient and the results do not justify the effort. Uranyl acetate, used as a general stain for sections, stains nucleic acids strongly, and this specific reaction is particularly marked at pH 4.8. (Adjust the pH of 2 % uranyl acetate to this value with dilute potassium hydroxide.) Post-staining with lead has a complementary effect, increasing the density of staining of the nucleic acids.

LIPIDS

Osmium tetroxide fixation gives a strong general staining of lipids. Casley-Smith (1963) has used osmium tetroxide to test specifically for unsaturated lipids. He used osmium for staining sections, staining for 1-3 hours in 2% aqueous osmium tetroxide. This stains unsaturated lipids very densely, saturated lipids less so. Because of the general staining properties of osmium, identical sections must be used as controls. Treat the controls with 2.5% bromine in carbon tetrachloride for one hour before the osmium staining. This completely removes the specific staining reaction of unsaturated lipids without affecting the general staining effect.

ENZYMES

Ultrastructural enzyme location depends upon forming a dense insoluble precipitate as the end product of a reaction catalysed by that particular enzyme. Techniques of this sort have been used to locate many different enzymes in animal tissue, but so far their use in plant cells has been very limited. Aldehyde fixation is essential for this work; osmium tetroxide very rapidly destroys enzyme activity.

Hall (1969, and personal communication) has studied phosphatase localization in plant tissue with success, using the Gomori reaction. This involves supplying the cells with substrate (an organic phosphate) and Pb++ ions. As soon as inorganic phosphate ions are liberated by the action of the enzyme, an electron dense precipitate of lead phosphate is formed at the enzyme site. Fixation of plant tissue is a problem, since if the fixation time is too long, phosphatase activity is reduced, while if it is too short the reagents do not penetrate into the cell. The best procedure seems to be to give a short fixation of 1 hour or less in 2% glutaraldehyde when studying phosphatases asssociated with the plasmalemma, but to give a longer fixation (2-4 hours) when studying phosphatases in the cytoplasm. The short fixation allows almost no penetration by the reagents, while a full fixation allows more or less total penetration of the cell at the expense of a very considerable loss of activity. Clearly there are certain pitfalls in this technique and, in particular, it is obviously impossible to make any quantitative comparison between phosphatase activities in different parts of the cell.

After fixation, wash the tissue segments thoroughly in several changes of ice-cold buffer. The staining medium consists of 2 mM ATP[1] (for ATPase) or 2 mM glycerophosphate (for alkaline phosphatase), 2 mM $Ca(NO_3)_2$, and 0.12% $Pb(NO_3)_2$, buffered to pH 7. Incubate the tissue portions for 30 minutes at 23°C. Wash the segments and then postfix in osmium tetroxide in the normal way.

POLYSACCHARIDES

The only cytochemical methods which have been developed specifically for plant material are those concerned with the localization of specific cell-wall polysaccharides. Several methods, giving different degrees of specificity, have been devised.

Ruthenium Red

Ruthenium red stains, in theory, carboxyl groups. In practice, this means that in plant material it exclusively stains pectins. It is used in very low concentrations $10^{-3}\%$ to $10^{-5}\%$, buffered to pH 7, either as a fixative, or as a postfixative after glutaraldehyde fixation. Incubate the tissues for at least 2 hours at 0°C.

Hydroxylamine-ferric chloride

Hydroxylamine in alkaline solution combines with methyl esters of pectins. The resulting compound (pectic hydroxamic acid) forms an insoluble complex with ferric chloride and this creates electron density. In light microscopy the scope of this technique can be increased by methylation with hot methanolic HCl to make all pectins stainable. This, however, tends to be too destructive to give satisfactory results in electron microscopy and, so far as ultrastructural work is concerned, the technique stains, therefore, only methyl esters of pectins. It has been suggested that esterification takes place as, or after, the pectins are incorporated in the wall. If this is so, this technique will only stain pectins already deposited in the wall, not those in, for example, Golgi vesicles.

Albersheim & Killias adapted the hydroxylamineferric chloride technique to the electron microscope, using it as a stain after fixation for tissue portions. However, the reagents are very corrosive and when used in this way cause an unacceptable amount of ultrastructural damage. The reaction can, though, be used with success to stain thin sections. *Nickel grids must be used, as ferric chloride dissolves copper.*

1. Alkaline hydroxylamine. Make up stock solutions consisting of

A. 7 g NaOH in 100 ml 25% ethanol.

B. 7 g hydroxylamine HCl in 100 ml 25% ethanol. Make up the working stain solution by mixing exactly equal volumes of A & B. Once mixed, the solution must be used immediately. Place the grids in this solution for 10 minutes (or longer, if necessary).

2. Acidification. Rinse the sections in distilled water or 25% ethanol and transfer them to 1 part conc. HCl + 5 parts 25% ethanol for one minute.

3. $FeCl_3$[1] staining. Place the sections for 2-3 minutes in 5 $FeCl_3$ in 0.05 HCl 25 % ethanol.

This schedule gives good results with Epikote embedded tissues. When using other resins it may be useful to vary the alcohol concentration in the solutions. Araldite, for example, is less permeable to water and better results may be obtained by increasing the alcohol content.

Silver hexamine

The periodic acid-Schiff reaction technique for locating polysaccharides has been adapted for electron microscopy by Pickett-Heaps. Periodic acid oxidises 1-2-glycol (and some other) groups to aldehydes, with which the Schiff reagent (in this case silver hexamine) reacts to give a silver precipitate. The specificities of the reaction in fixed plant cells are not very well understood and different fixation methods and pretreatments give different staining patterns. The treatment can cause tissue damage and the resolution is limited by granularity in the silver deposit. On the other hand, both the specificity and the staining are good and the method offers considerable promise when it is better understood.

OTHER CYTOCHEMICAL METHODS

Ferritin-antibody

Antigens can be located by using an appropriate antibody, made visible in the electron microscope by covalently coupling ferritin molecules to the molecule of the antibody. This technique has been used to study surface antigens in bacteria and animal tissue and to a certain extent to study internal antigens, although it is difficult to make cells permeable to antigens without severely damaging them. Details of the basic procedure are given by Singer & Schick and of its application to micro-organisms by Mott.

Enzyme digestion

Specific chemical components in cells can be identified by selective extraction, using enzymes or acid hydrolysis. As applied to bulk tissue, this method suffers from the risk of causing gross ultrastructural damage, but Leduc & Bernhard and Leduc,

Marinozzi & Bernhard have described a range of techniques using thin sections embedded in water-soluble media. These give excellent structural preservation without preventing selective digestion, and have considerable promise.

NEGATIVE STAINING

The Theory of Negative Staining

There are three principal ways of deriving information from an object. Shadowing and positive staining are described elsewhere. Negative staining makes possible the resolution of, for example, protein molecules and assemblies of lipid micelles and justifies the use of instrument magnifications between 50,000 and 100,000 times. The chosen particle is surrounded and embedded in electron dense material. The electron dense material does not

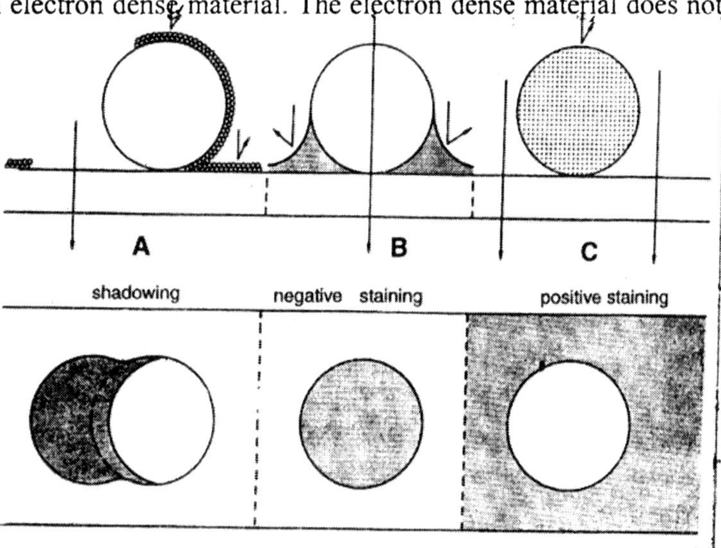

Figure 5.2

remain particulate, but dries to a glass-like state. What happens on the boundary between the negative stain and the object is even less well understood than the reactions taking place in positive staining. The stain does not generally penetrate deeply into the specimen, but renders it visible by the creation of an abrupt boundary between a low and high contrast subject. Local pH changes, alterations in the concentration and surface tension effects as the stain dries around the object make it very difficult to predict what will happen to a given object at a certain pH or concentration. Moreover, as Glauert & Lucy showed, negative staining may produce very different images with only small changes in the preparative procedure.

Negative staining of small particles e.g. viruses and bacteriophage

Purify the virus preparation. Use a concentration of 10^7 to 10^8 particles per ml. Using a pasteur pipette, place two or three drops of the virus suspension on a holey-Formvar coated 400 mesh grid. Drain the grid by holding it between the tips of a pair of forceps, insert a pasteur pipette at an angle of 45° and drip two or three drops of the chosen negative stain across the grid so that the majority drains straight off. Drain off again on a fluffless filter paper. Examine in the microscope. If the negative stain is insufficient for the amount of virus present run a few more drops across the specimen.

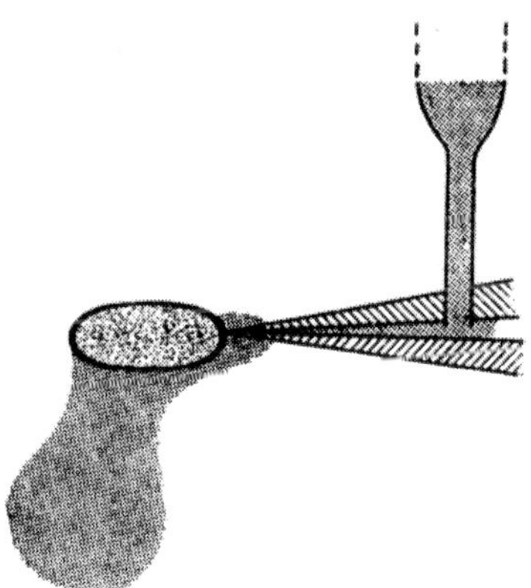

Figure 5.3

Horne suggests that if the virus and stain do not spread properly on the grid between 0.005 % -0.05 bovine serum albumin or glycerine or propylene glycol may be added to the particle preparation. A list of negative stains and the pH ranges which have been found successful is given below. Different stains may behave very differently. For example, potassium phosphotu-ngstate penetrates into the surface of many viruses or down the hole in rod-like viruses such as TMV, whereas ammonium molybdate does not normally do so. Uranyl formate demonstrates the helical construction of TMV rods, whereas no other negative stain adequately does so. Therefore use, on a virus or any other specimen, a range of negative stains and a range of pH levels.

TABLE 5.3 : SOME NEGATIVE STAINS AND THEIR RESPECTIVE PH RANGES

Negative stain	*Final conc.*	*pH*	*buffer*
Potassium phospho tungstate	2°/₀	6.0-7.0	ammonium acetate
Uranyl acetate	0.5%-2%	4.0-5.2	none
Uranyl formate	0.5%–1%	4.0-5.2	none (unstable, does not keep)
Ammonium molybdate	2.0%-3%	6.0-8.0	ammonium acetate
Lithium tungstate	2%	6.0-7.5	" "
Sodium tungstoborate	2%	5.5-7.0	none

Standardise the length of time the virus preparation is exposed to the negative stain before being examined in the microscope. Some stains may continue to penetrate into the protein coats as they slowly take up moisture from the atmosphere. Never attempt to store either mixtures of virus and stain or prepared grids. Virus structure breaks down very rapidly under these conditions.

Negative staining of large particles, e.g. bacteria, microtubules, fragments of mitochondria

Purify the bacterial preparation. Use a concentration of bacteria of about 10^5 particles per ml. Drip a little of the bacterial suspension onto a Formvar-coated 400 mesh grid. Drain and dry for two minutes. Drip a few drops of the negative stain across the surface of the grid. 1% w/v tungstoboric acid gives good results as a stain for bacteria. Drain immediately, dry and examine.

6

MOUNTING

A mounting medium is a substance in which tissues are immersed for examination under the microscope. The same name should not be applied to adhesives that merely stick sections to slides and do not enclose the tissues at the moment of examination. Mounting is necessarily the last process to which tissues are subjected in the making of a microscopical preparation.

Solid, unshrinkable objects may be mounted dry. If so, air is to be regarded as the mounting medium. In electron-microscopy sections are examined dry, in a vacuum; if the embedding medium has been removed, the tissue constituents are exposed directly to the electron beam. With these exceptions, tissues are always mounted in media that are fluid at the outset, though many solidify later. The liquid may be a single pure substance, a mixture of liquids, or a solution of one or more solids in a liquid; it must be optically homogeneous and stable. Volatility is allowable, because the edge of the mount can be sealed. The mounting medium must fill even the minutest spaces formed within cells by coagulant fixatives, for otherwise different parts of the specimen would be permeated by fluids of different refractive indices, and the image would therefore be confused. The mounting medium must be transparent and nearly or quite colourless. It must be capable of wetting the various tissue-constituents, especially proteins (that is to say, it must be able to lie against them without leaving any intervening spaces). It must not dissolve or corrode the tissue-constituents that are to be studied, though solvents of particular constituents (such as lipids) are permissible if the intention is to study the insoluble remainder of the tissue. It must not support the growth of moulds or bacteria.

The refractive index of the medium must be related to the means adopted to obtain contrast in the image produced by the microscope. An example will make this clear. Let us imagine that a piece of tissue has been fixed by formaldehyde, and that all lipids, carbohydrates, and nucleic acids have been dissolved out of it. The specimen will consist mainly of protein, and this, in its fixed condition, will generally have a refractive index of about 1.536. If the fixed tissue is now thoroughly permeated by a colourless, transparent fluid of the same or very nearly the same refractive index, such as methyl salicylate, the whole object disappears and no image of it can be obtained by the microscope.

A fluid that has this effect is often called a 'clearing agent', but the term is misleading. The fixed proteins of the tissue are transparent and colourless, and require no clearing. To render the tissue as a whole transparent, it is only necessary to fill all the spaces intervening between the proteins with a transparent, colourless fluid of the same refractive index. There is then no diffraction of light at the surfaces of the protein, and therefore no image of it can be formed. The protein is unaffected. Methyl salicylate no more 'clears' the proteins than a transparent, colourless fluid of the same refractive index as glass 'clears' a glass object immersed in it.

To produce contrast and thus prevent invisibility caused in the way just described, one may dye the tissue and then mount in the same medium as before. The object becomes visible because light of certain wave-lengths is reduced in intensity by the dye attached to the proteins, but passes freely through the mounting medium. The fact that the latter has the same refractive index as the proteins is now helpful, because the complete homogeneity of protein and medium in this respect is favourable to the production of an optically perfect image.

Alternatively one may leave the object undyed, but render it visible by mounting it in a fluid that differs slightly from the proteins in refractive index. The specimen should then be examined by phase-contrast microscopy. To get the best effect, the difference in refractive indices should be small. The retardation or advancement of the light that passes through the object in relation to that which passes only through the mounting medium should not greatly exceed a quarter of a wave-length of the light used for examination, if phase-contrast is to be applied effectively.

A transparent object is rendered visible by direct ('ordinary') microscopy if it is mounted in a medium of much lower or higher refractive index than its own. This is convenient if nothing more is needed than a low-power view of a whole mount; but wherever the object comes up against the medium there will necessarily be a Becke line in the microscopical image, and this is a misleading appearance, since it does not represent anything that was present in the object. The resolution of minute detail is not possible in such circumstances.

Mounting media may be classified as hydrophil and hydrophobe. The hydrophil media are those that contain water or are miscible with water. Sections or whole mounts may be transferred from water to hydrophil media without thenecessity to use an intermediary liquid not contained in the medium. It is reasonable, however, to introduce the mounting medium gradually to the tissue. Thus, in mounting in glycerol or Farrants's medium one may pass the section or other object through a mixture of water and glycerol before mounting in the medium itself. The two substances may conveniently be mixed in the proportion in which they occur in Farrants's medium (50 g = 41.7 ml of glycerol with 100 ml of distilled water).

Hydrophobe media require the dehydration of the object, since they are not miscible with water. It is usual to pass tissues through a graded series of ethanols (often 70%, 90%, absolute). Tissues that have been dehydrated by ethanol in the course of embedding have undergone all the shrinkage of *which* they are capable, and no harm could come from direct passage of sections from water to absolute ethanol. However, the absolute ethanol would soon be diluted if this were done repeatedly, and similarly the 90% ethanol would soon lose its strength if direct passage were made from water to ethanol of this concentration. The closer the grading of the ethanols, the more accurately they will maintain their concentration, but the more trouble will be involved in dehydration. The usual series provides a good compromise. The dehydrated section is usually transferred to the solvent contained in the mounting medium, and from this to the mounting medium itself. Some hydrophobe media, however, contain no solvent. Methyl salicylate is an example. The tissue is transferred directly from absolute ethanol to such media.

Both hydrophil and hydrophobe mounting media are further subdivisible into those that are adhesive (and thus bind the cover-slip to the slide) and those that are not. If a preparation is to be kept permanently, it is almost essential that the coverslip should be bound to the slide. The easiest way of achieving this is to use an adhesive mounting medium.

If perfect apposition can be obtained between two solid objects, they will ordinarily adhere, because their molecules will be drawn towards one another by the same forces that hold the molecules together in the objects themselves. There is no question, however, of causing coverslip and slide to adhere in this way, partly because their surfaces cannot be made flat enough for perfect apposition on the molecular scale, partly because a section or other microscopical object must necessarily intervene. Some degree of adhesion will be achieved if a fluid intervenes, provided that the fluid is able to wet both the surfaces concerned. It is a fortunate circumstance that glass can be wetted both by water and by the solvents used in hydrophobe mounting media. The adhesion will be very poor, however, unless the liquid not only adheres to the glass but also coheres within itself. In other words, it must be viscous, or else be capable of actual conversion into a solid. The presence of long molecules in the liquid will thus favour adhesion. So that the surfaces may be easily wetted, it is desirable that the adhesive should be at first of fairly low viscosity. The necessary increase in viscosity or conversion to solid may be achieved either by cooling or by the evaporation of a solvent.

The adhesive hydrophobe mounting media, such as DPX are hardened by evaporation of the solvent, which is usually xylene. The commercial product that is sold under this name is a mixture of three isomeric dimethyl benzenes with ethyl benzene. It has the percentage composition shown here, whether it is derived

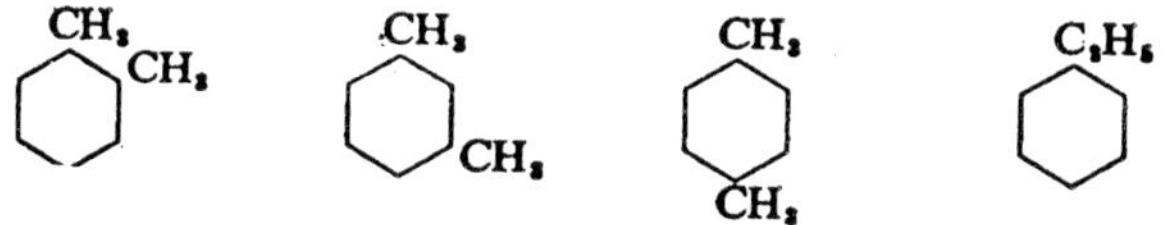

o-xylene,23% m-xylene, 43% p-xylene, 19% ethyl benzene, 15%

The components of xylene

from petroleum or coal. Xylene is chosen as a solvent for hydrophobe mounting media because it evaporates sufficiently slowly to allow the coverslip to be adjusted at leisure. Hardening is usually hastened by putting the slide on a warm plate after the coverslip has been arranged in position.

Hydrophil media are used when low refractive indices are required, or when it is desired to avoid dehydration at all stages. Most hydrophobe media have high or rather high indices, approximating to those of proteins that have been fixed by formaldehyde or by coagulant fixatives. These hydrophobe media generally give more transparent preparations than hydrophil ones.

Certain non-adhesive media present the advantage that the final refractive index is known, because there is (or should be) no change in composition if evaporation occurs.

One cannot state the effective refractive indices of those adhesive media that rely for adhesion on the evaporation of a volatile component, because the exact amount of this component (if any) that remains in the preparation at any particular time cannot be controlled. One can only state the extreme limits of refractive index: that is to say, the index of the complete medium, ready for use, and the index of the medium from which the volatile component has been completely driven off. It is unlikely that this component is quite eliminated from ordinary microscopical preparations.

Since high refractive index and firm adhesion of the coverslip are desirable in most cases, adhesive hydrophobe media are more often used than the others.

When non-adhesive media are used, it is generally desirable to stick the coverslip to the slide by the use of a seal, which prevents movement and also the evaporation of any volatile constituent.

Four mounting media will be described here. They are representative of the very large number that exist. They have been chosen because they are simple in composition and therefore demonstrate well the principles involved in the process of mounting. The selected examples are these:

HYDROPHIL

non-adhesive (fluid mount)	glycerol (see below)
adhesive	Farrants's medium

HYDROPHOBE

non-adhesive (fluid mount)	methyl salicylate
adhesive	DPX

Glycerol (glycerine) is a familiar non-adhesive hydrophil medium. It is useful in studies of lipids, which are preserved whether fixed or not. Lysochromes maintain their colour well in this medium, but certain dyes, including mordanted haematein, are not perfectly stable. Glycerol is very valuable as a mounting medium for sections of tissues that have been embedded in butyl methacrylate and are intended for study, without dyeing, by phase-contrast microscopy. When preparing tissues for electron-microscopy it is desirable to cut a few sections at about 3μ and to examine these by phase-contrast. This shows at once whether the tissue is well enough fixed to make it worth while to cut thin sections and mount them on grids for electronmicroscopy. It also helps in the interpretation of electronmicrographs, because the thickness of the section enables one to focus up and down and thus obtain a three-dimensional view. Material that has been fixed with osmium tetroxide, embedded in methacrylate, sectioned at about 3μ, and mounted in diluted glycerol, gives an extraordinarily lifelike appearance under the microscope. A few authors, have studied methacrylate sections by phase-contrast, but not enough has been done in this promising field.

It is generally supposed that the fixed proteins of the cell have a refractive index of about 1.536. This supposition rests on observations made with tissues fixed in formaldehyde solution and embedded in paraffin. There is reason to believe that the proteins of cells that have been fixed by osmium tetroxide and embedded in butyl methacrylate have a considerably lower index than this. It is desirable that the index of the mounting medium should be somewhat lower again, if undyed sections are to be examined by phase-contrast. The ground cytoplasm will then appear very pale grey by positive contrast, while the mitochondria and certain other cytoplasmic inclusions will be dark grey or black. A mounting medium of refractive index 1.46 has been recommended. Media with even lower indices, such as 1.44 or 1.42, seem to give still better results. The refractive index of pure glycerol is 1.474. By mixture with water, any lower refractive index exceeding that of water itself (1.333) can be obtained. The proportions of glycerol and water that give various refractive indices are shown in published tables. To make a fluid of index 1.44, 78 g of glycerol are mixed with 22 ml of distilled water; of index 1.42, 65 g of glycerol with 35 ml of water.

By the addition of cadmium chloride, the refractive index of glycerol may be increased to 1.54 at saturation. The salt imparts a pale yellow tint to this otherwise colourless medium.

Farrants's medium is a good example of an adhesive, hydrophil medium, much used in studies of lipids. It is not nearly so good as diluted glycerine for the study of undyed methacrylate sections by phase-contrast

Several different formulae have been published, all containing the same three principal ingredients. The following works well.

To 40 g of gum arabic (preferably powdered), add 40 ml of distilled water. When the gum has dissolved, add 20 g of glycerol. A piece of camphor may be put in the medium to act as a disinfectant.

The pH of this fluid is 4.1, whether camphor be added or not.

True gums are sticky, water-soluble substances that exude

through cracks in the bark of certain trees. Gum arabic (acacia gum) is derived from species of *Acacia* (Leguminosae), especially *A. senegal*, a native of the Sudan and other parts of tropical Africa. The gum is taken from both wild and cultivated trees. Its name refers to the fact that it was first imported into Europe from Arabian ports. The gum consists of the calcium salt of arabic acid, which owes its acidity to the carboxyl group of glucuronic acid. Hydrolysis yields arabinose, rhamnose, and galactose, in addition

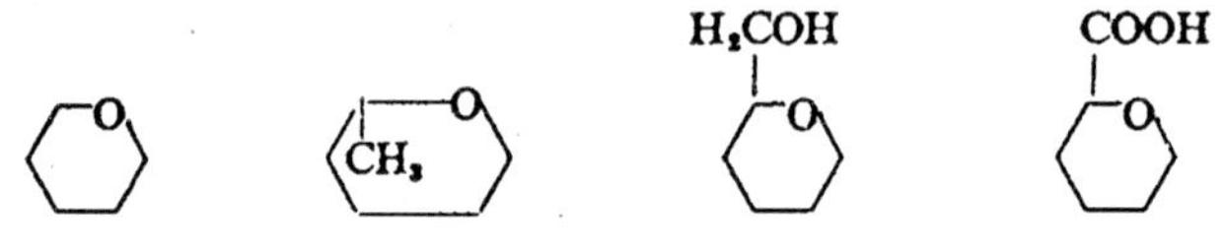

arabinose *rhamnose* *galactose* *glucuronic acid*

Skeleton formulae of the hydrolytic products of gum arabic

to glucuronic acid. The exact way in which these are bound together in the natural product is not known, but the evidence suggests a branched chain, in which the components are not arranged in a regularly repeated order. The molecular weight is about 240,000. The shape and size of the molecule determine its high viscosity in aqueous solution.

The refractive index of the medium made up according to the formula given above is 1.423 (or 1.424 if saturated with camphor). It is best to drive off most of the water (and thus increase the refractive index) by leaving the slide overnight in the paraffin oven, after applying the coverslip. The latter adheres firmly.

It has been stated in print that 'The ordinary media of the Farrants' type have an index of refraction just over 1.3. This statement cannot be correct, whichever formula be followed; for the refractive index of water itself is 1.333, while those of the other constituents are very much higher.

The glycerol in Farrants's medium prevents the gum from cracking when the water is driven off. It also helps to make the tissue transparent, because it readily penetrates the minute spaces in them, while the large molecules of the gum diffuse slowly. The uneven distribution through the tissue of the two substances of high refractive index in this medium is probably the cause of the imperfection of the phase-contrast image.

Methyl salicylate. Non-adhesive hydrophobe media are very seldom used, partly because most of the suitable substances are inconveniently volatile, partly because it is so easy to prepare hydrophobe media that will stick firmly to glass. Methyl salicylate, however, has certain virtues. It is colourless and miscible in all

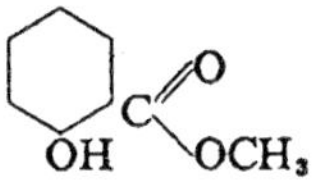

Methyl salicylate

proportions with absolute ethanol. It gives excellent transparency to most preparations, because its refractive index (1.537) is very close to that of proteins fixed in the usual ways. Since it is a single substance, there is no question of change of refractive index by differential evaporation of components. Basic, acid, and mordanted dyes seem to maintain their colours well in it.

DPX. It has already been remarked that adhesive hydrophobe mounting media are more commonly used than any others. They are valuable for making permanent preparations of dyed objects. Dyes maintain their colours well in DPX, which possesses the further advantage of being itself perfectly colourless. The letters DPX are the initials of Distrene-Plasticizer-Xylene.

The most familiar of all media in this group is Canada balsam, which was described in earlier editions of this book. Although it has great virtues, it is omitted from the present edition, partly because it has a tendency to bleach basic dyes, partly because it is very complex in composition (there are at least 7 constituents in addition to the solvent).

In adhesive hydrophobe media the essential constituent is a 'resin', in the wide sense of that term: that is to say, a solid, amorphous organic compound, insoluble in water, having no definite melting-point (because composed of molecules of varying lengths). The natural resins that exude from the bark of certain trees are' oxidation-products of terpenes, but plastics having similar physical characters nowadays receive the same name. The particular plastic that gives high refractive index and adhesiveness to DPX is polystyrene, a polymer having the same general character as acrylic acid, but differing in the substitution of a phenyl group for the carboxyl of the acid. The polymer that is used in DPX has rather less than 800 segmers in most of its

A segmer of acrylic acid A segmer of polystyrene

molecules. It is known by the trade name of 'distrene 80'. It is a colourless solid, freely soluble in xylene.

If a section or other object be mounted in a solution of distrene 80 in xylene, the polymer retracts under the edges of the coverslip as the solvent evaporates. This defect is mitigated or overcome by the addition of a plasticizer to the medium. Typical plasticizers are non-volatile solvents of the plastic on which they act. They are supposed to associate themselves with reactive groups in the polymer, which would otherwise attach themselves to other reactive groups, and thus link the long molecules into a rigid meshwork; but they do not combine with the polymer so as to form a new substance. Thus they make the polymer softer than it would otherwise be, and oppose its contraction when any volatile solvent evaporates. The plasticizer in DPX is tri-p-tolyl phosphate (often called tricresyl phosphate); that is to say, an ester of phosphoric acid, in which three p-tolyl radicles have replaced three hydrogens.

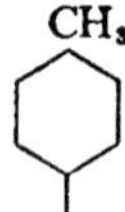

The p-tolyl radicle

This plasticizer is a colourless, non-volatile liquid of high refractive index (about 1.56).

To prepare DPX, add to 100 ml of xylene 18.75 ml of tri-p-tolyl phosphate, and then 25 g of distrene 80. The latter is obtainable from Messrs Honeywill and Stein, 21 St James's Square, London, S.W.1.

The plasticizer is irritating to the skin and contact should be avoided.

The refractive index of DPX, while it still contains the original amount of xylene, is 1.532. Since the index of xylene is only about 1.496, that of the medium must increase as drying proceeds. The final index is rather higher than that of the fixed proteins of ordinary microscopical preparations. If all the xylene is deliberately driven off from DPX, the resultant substance (distrene + plasticizer) is a solid, melting above 100°C. Although the plasticizer is a solvent for distrene, not nearly enough of it is present to make a solution.

In whole mounts there is some tendency to retraction under the coverslip, despite the presence of the plasticizer.

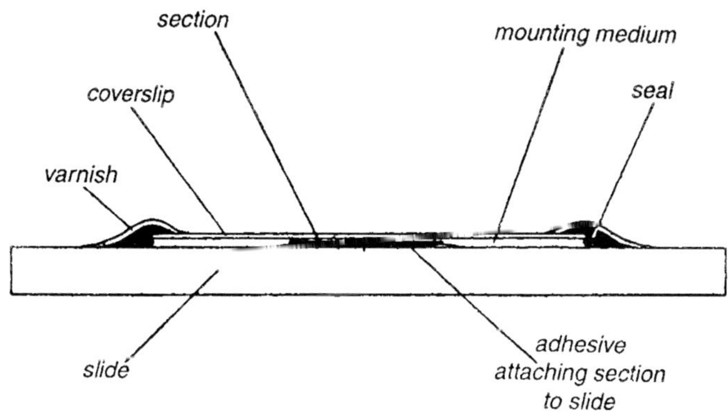

Figure 6.1 : *Diagram of a finished microscopical slide in sectional view, showing adhesives. Teh seal (and varnish) are only needed if the mounting medium is fluid.*

Permanent preparations can be made of objects mounted in non-adhesive media, by sticking the periphery of the upper surface of the coverslip to the slide. The gelatine gel used for embedding is a good adhesive for this purpose. It melts readily in an incubator at 37°C. The slide is dried with a cloth up to the edge of the coverslip, and the melted gel applied with a paint brush. It hardens at first by becoming a gel again, and subsequently by evaporation of the water. In a day or two it becomes so hard that it can scarcely be marked with the finger-nail. Both hydrophil and hydrophobe mounting media can be sealed with gelatine.

As an extra precaution against evaporation of the mounting medium, it is wise to varnish the hardened gelatine and to extend the varnish beyond the gelatine on to the upper surfaces of the coverslip and slide. The varnish chosen for this purpose should be one that makes perfect contact with glass surfaces and is not readily softened or dissolved by the various fluids used as immersion-oils. Gold size is convenient for the purpose. This is a substance used by gilders in the preparation of surfaces to receive gold foil. It consists essentially of an oleo-resin dissolved with turpentine and linseed oil. The oleo-resin (or resin dissolved in an essential oil) is often wrongly called 'gum' animi. It exudes from the bark of a leguminous tree, *Hymenaea courbaril*, a native of the West Indies. Two separate processes are involved in the drying of this varnish: first, the evaporation of the essential oils; secondly, the oxidation by atmospheric oxygen of the linoleic and other unsaturated fatty acids contained in the linseed oil. This oxidation is helped by heating the oil with lead monoxide in the preparation of the varnish. The hardening occupies some days after the varnish has been applied.

All finished microscopical preparations of organisms and their parts, other than those of living cells still lying in the fluid that bathed them in their natural environment, are to be regarded as products of the reaction between the living tissues and the various media in which they have successively been soaked. The reactionproducts, which are what we study under the microscope, are only informative about organisms and their life-processes if some knowledge is available about the reactions involved; and this presupposes knowledge of the reagents. It has been the purpose of this book to supply information about the reagents that are used in some of the simplest processes of routine microtechnique, and about their reactions with tissue-constituents.

If an unknown tissue or cell be acted upon by an unknown reagent, no useful information can be obtained. Cytologists and histologists should be as loath to use secret reagents as doctors are to use secret medicines. Quite a lot of fancifully named embedding and mounting media of unstated chemical composition are in use today, despite the fact that their reactions with tissue-constituents cannot be understood while their formulae remain secret. Even the field of electron-microscopy, in which a scientific outlook might have been expected, has been invaded by makers of secret mixtures. For instance, we are asked to use 'hardeners' and 'accelerators' to solidify an embedding medium, without being told what these reagents may be. We cannot interpret the resultant electron-micrographs satisfactorily without this information, because it is necessary to know whether any tissue-constituents are likely to have been dissolved or affected in some other way. Reagents of all kinds used in microtechnique should be prepared in the biological laboratory, or obtained from reliable manufacturers who announce the composition of their products. The only possible exceptions are such substances as varnishes, which do not come into contact with the tissues.

7

SPECIAL TECHNIQUES

PLANT VIRUS PREPARATION

The following methods will give a quick and reliable test for the presence or absence of most viruses in plant tissue. More sophisticated cleaning and concentrating techniques will be found in Maramorosch & Koprowski, Vol 2.

Epidermal peel method

Peel away with a pair of fine forceps, a piece of the lower (abaxial) epidermis from a freshly gathered leaf. Dab the torn surface once or twice on to a drop of 2% neutral buffered phosphotungstic acid supported on a coated grid. Formvar, carbon-coated Formvar or carbon films can all be used. Drain away the excess PTA from the edge of the grid with a fluffless filter paper. Examine in the microscope. If the leaf will not peel easily, pull away a small fragment, crush between the tips of the forceps and dab on to the PTA as before.

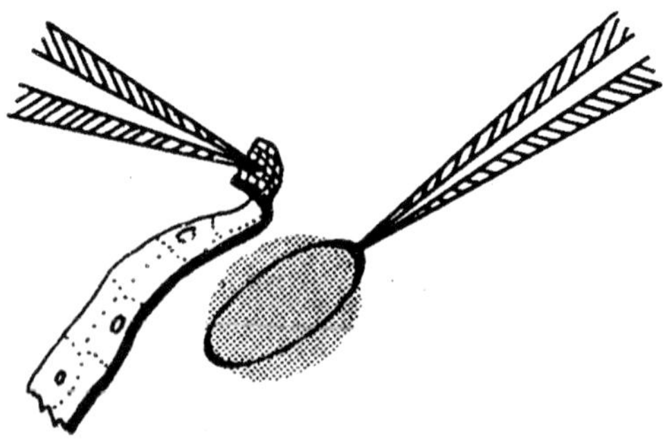

Figure 7.1

Extrusion method

This method is for use with heavily thickened leaves, conifer needles or stems. Cut either a third order vein of the leaf or across the needle or stem and dribble or squeeze the exudate out on to a PTA drop as above. Use hydraulic pressure or a converted grease gun to force out a visible exudate from old stems. PTA even at neutral concentrations may not give satisfactory preparations with some viruses. If this is suspected, use a drop of 2% neutral buffered phosphomolybdic acid, or unbuffered 2% solutions of uranyl acetate, formate or oxalate.

VISIBLE TRACERS FOR THE ELECTRON MICROSCOPE

Ferritin

The globular protein ferritin has been used in electron microscopy for some time to study the passage of material through blood capillary surfaces and other tissues in which pinocytosis is found in animals (Haggis); the molecule can be seen in the microscope because of its iron core which is 5.5 nm in diameter. Barton used suspensions of ferritin with limited success to study uptake by root tips of *Phaseolus vulgaris*. Roots were placed in 0.25 and 0.5% suspensions of ferritin in water for 2 to 24 hours. Most of the ferritin was observed in the cell walls, with some apparently bound to the plasmalemma. No conclusive evidence of cytoplasmic incorporation was found. However Cocking found that ferritin in the form of a 5% water suspension was taken up by isolated tomato fruit protoplasts by a mechanism involving pinocytosis.

Ferritin should be a safe non-toxic marker to use in tracing work, as it occurs naturally in plant cell organelles according to Robards & Robinson and others. It is thought to be a possible method of storing iron in a form not toxic to the cell. Use ferritin free of cadmium for tracer work; cadmium is a very common and toxic contaminant. Do not use potassium perman-ganate as a fixative in conjunction with ferritin, as it tends to destroy any ferritin present.

Tobacco mosaic virus

Cocking demonstrated that isolated tomato fruit protoplasts will take up TMV particles from suspension by pinocytosis. The plasmalemma invaginates where TMV particles are attached, and the resulting pinocytotic invagination may include free particles as well. The usefulness of this as a marker is uncertain, since although TMV (17 nm in diameter) is easily seen and identified in the cell, it may itself be inducing the pinocytotic uptake; if so the pinocytosis observed may not be a natural function of the cell.

Gold colloid

Gold colloid particles of a diameter of 4 nm, prepared using chlorauric acid and potassium carbonate, have been used by Gaff, Chambers & Markus to trace water movement in plant cell walls. The particles are able to move between cell wall fibrils with sufficient ease to be carried along in the transpiration stream. The cut ends of leaves of *Helxine soleirolii* were placed in a 0.12% gold sol and the leaves allowed to transpire for varying lengths of time. Gold particles were observed to have been taken up in the wall, but not by the cytoplasm. An odd feature of the study was that of the several plants used, only *H. soleirolii* could be induced to take up the gold particles.

FREEZE-ETCHING

Thin-sectioning methods cannot give a surface view of membranes or organelles. This limits the information that they can give about the interior of the cell. The freeze-etch technique allows us to view these surfaces by shadowed replicas similar to those used for external surfaces.

First the specimen is frozen sufficiently rapidly to prevent ice-crystal formation. It is then, under high vacuum, fractured by a razor blade. Water vapour is allowed to sublime from the exposed, still frozen, surface. A selfshadowed replica is then made of this surface.

The technique in practice demands a high degree of skill and the equipment is both complex and expensive. It is unlikely, therefore, that anyone will have access to the equipment without being given comprehensive practical instruction, and the notes following are intended to be used in conjunction with such instruction.

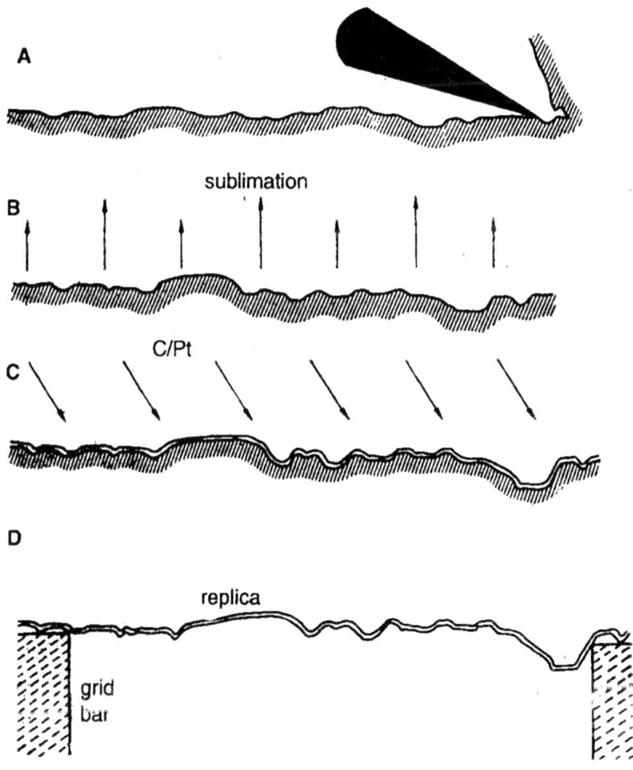

Figure 7.2

Specimen preparation

Vacuolate plant cells, with their high water content, are much harder to freeze successfully than yeast or animal cells. It is essential to use 20% glycerol as an antifreeze agent and this high concentration is normally enough to cause plasmolysis. Most workers therefore fix the tissues first.

Fix the tissue in 2 % glutaraldehyde at room temperature for 20 minutes. If this does not give satisfactory preservation, use a full fixation time of 2 hours. Transfer to 20 % glycerol in water for at least 24 hours. Place a small piece of treated tissue inside the raised rim on a copper disc. For liquid cultures use a flat copper disc with the surface scratched to improve adhesion, and place a small drop of the solution in the centre of the disc.

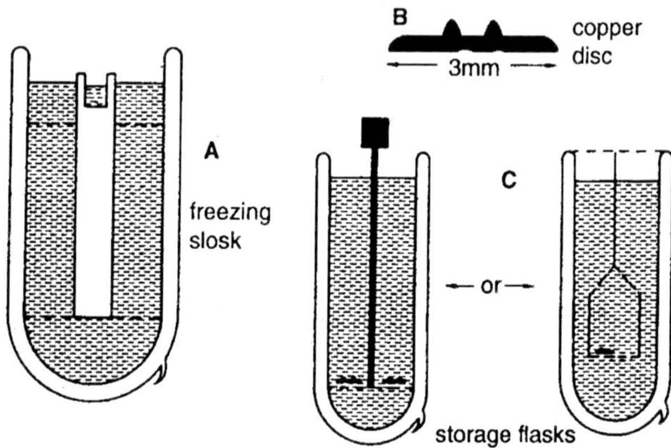

Figure 7.3

A freezing flask is normally supplied with the freeze-etching equipment. Fill the outer container with liquid nitrogen 25 Run Freon 12[25] or 22[25] gently into the small metal inner container. The gas will liquefy and collect in the container and finally solidify. Melt the centre of the frozen Freon with a metal rod at room temperature and plunge the specimen in, holding the disc with a pair of very fine insulated forceps. After a few seconds transfer the specimen quickly to a storage flask of liquid nitrogen. Figure 20C shows two types of storage flask. In the first the disc is simply dropped in and can be removed by raising the perforated plate. The second uses an aluminium can with a perforated lid. Fill this with liquid nitrogen, drop in the disc, attach the lid, and lower the whole into a Dewar flask of liquid nitrogen.

Cutting and etching

Before inserting the specimen, cool the stage to −150°C so that the specimen will remain frozen when it is placed in position. Keep the bell-jar under vacuum while the stage is cooling, or ice crystals will form.

When the stage has reached –150°C close the vacuum valves, raise the bell-jar and, as quickly as possible, paint the stage with liquid Freon and place the specimen disc in position. Cool the retaining cap in liquid nitrogen and screw it on. It is essential to carry out these operations in a minimum of time, both to prevent the temperature of the specimen rising and to minimise icing-up of the stage. Replace the bell-jar, start pumping it out and set the cooling controls to cool the knife, and bring the stage to –100°C.

The vacuum must reach 5×10^{-6} mm Hg or better before the specimen is cut. The stage must be at –100°C and the knife at the temperature of liquid nitrogen. When these conditions have been reached, cutting can begin. It is important always to move the knife quickly across the specimen when cutting; a slow cutting speed will dislodge the specimen from its holder. Move the knife rapidly round its cutting path, at the same time lowering it gently until the specimen is being cut. Then use the fine advance, or advance very gently with the manual control, until a good surface has been prepared. Look for a mirror-like reflection from a good surface.

Some workers prefer to make the first cuts before the knife is cooled. Advance the knife gently, as before, until a mirror-like surface has been prepared. Then cool the knife and make a fev more cuts, advancing gently, until the surface of the specimen becomes very slightly cloudy. The surface is then ready for etching. This method reduces the danger of dislodging the specimen and produces a more level surface, at the expense of increasing the risk of recrystallization.

To etch the surface retract the knife on its coarse advance and place it over the specimen, where it will act as a cold trap for the sublimed water vapour. Allow sublimation to take place at –100°C for a short while; 30-60 seconds will normally be enough, but up to 2 to 3 minutes can be used if necessary.

Remove the knife and evaporate the platinum-carbon mixture on to the surface. A piece of white card cut to fit round the stage makes it easier to check that the right amount is deposited. Next evaporate a thin coat of carbon on to the specimen from the electrodes above it, and the replica is complete. Place the knife over the specimen again. This will help to reduce thermal shock, which could crack the specimen and disrupt the replica when the vacuum chamber is opened. Raise the bell-jar, remove the specimen and float the replica on to the surface of a 20% glycerol solution at room temperature, or directly on to the cellulase solution if this is available.

As soon as air is admitted to the vacuum chamber all the cooled parts will become coated with frost. Before the apparatus can be used again the stage and knife assemblies must reach room temperature. The heating elements in the apparatus will help, but if the equipment is wanted again immediately turn a fan heater on to the stage and knife. When the frost has melted, remove any drops of water by drying with tissues. The electrode holders and shields must also be cleaned, the carbon electrodes sharpened and the platinum-carbon electrodes re-made. Align the electrode assemblies when they are put back so that the hole in each shield is aimed at the stage. Align the off-set platinumcarbon electrode by eye. To line up the carbon electrode exactly above the specimen, pass a small current-just sufficient to give a noticeable glow-through the electrodes, and move the shield until the glow falls on the stage.

Cleaning replicas

Pretreatment with cellulase to soften the cellulose of plant materials is almost essential before chemical cleaning. First, therefore, float the replicas on a cellulase solution for two days. Then transfer the replicas, using a platinum loop, to chemical cleaning agents, such as household bleach followed by sulphuric or chromic acid. Float the replicas for 24 hours on each. Finally wash in distilled water before examining.

SPECIMEN SUPPORT FILMS

Even with the large range of specimen support grids now available, it is sometimes still necessary to coat the grid with a support film to study certain materials, e.g. plant virus preparations, bacteria and isolated ribosomes. Similarly very thin sections (less than 50 nm) supported on large mesh grids (75 or 100 meshes/inch) for making mosaics of large areas of plant tissue will require some additional support, otherwise the sections become badly distorted and often fall through the grid bars.

As the support film must be rigid, structureless, and yet thin enough to have a high electron transparency, the range of suitable materials is restricted. Also the film must be stable under high vacuum and electron bombardment. Films are normally made from evaporated carbon or silica, or a plastic. Formvar (polyvinyl formaldehyde) and collodion (nitrocellulose) are the commonest plastics used.

Collodion films

Make a 2% w/v solution of collodion in amyl acetate. Fill a dish of 20-30 cm diameter with distilled water and place on the bottom of the dish a large circle of fine wire gauze. Place on the gauze a quantity of grids. Allow 2 or 3 drops of the collodion solution to fall gently from a pasteur pipette on to the surface of the water. When the film has set, remove it using a mounted needle; this first film cleans the surface of the water and is discarded. Alternatively, draw a sheet of completely fluffless paper (certain makes of toilet paper are very suitable) across the surface. A second film is cast on the water as before and the distilled water is gently syphoned off. The gauze and coated grids are allowed to dry before the grids are detached from the gauze.

Alternative method

Clean the water surface as before and prepare a fresh film. Gently place grids on the film. Take a clean glass bottle smaller than the width of the casting tray and with a rolling motion start at one end of the film and pick it up on the side of the bottle. Instead of a bottle the film may be picked up by dropping a piece of fluffless paper onto the film and carefully lifting from the surface. Allow the paper to dry. The grids can be picked off as required.

Formvar films

Make a solution of Formvar in chloroform or ethylene dichloride, 0.2%-0.5% w/v depending on the thickness of film needed. Wipe a glass microscope slide with a clean cloth. Dip it into the Formvar solution and drain vertically on to filter paper under cover until it is dry. Score the film 1-2 mm from the edge of the slide with a razor blade corner. Lower the slide slowly at a shallow angle into a dish of distilled water. Surface tension effects should lift the film off the slide and float it on the surface of the water. It may help to breathe on the film before immersion to ease its detachment and make it more visible on the water surface. The film can be picked up on grids as for collodion.

Collodion and Formvar films are not suitable for high resolution work as electron bombardment causes some distortion and a slight amount of film decomposition. For high resolution work, carbon or silica films are recommended.

Carbon films

Method 1

Smear a clean glass microscope slides with a little detergent (e.g. Teepol). Polish off with a clean cloth. Place the slide under the bell jar of a vacuum evaporation unit. Next to the slide place a small piece of glazed white porcelain with a drop of high

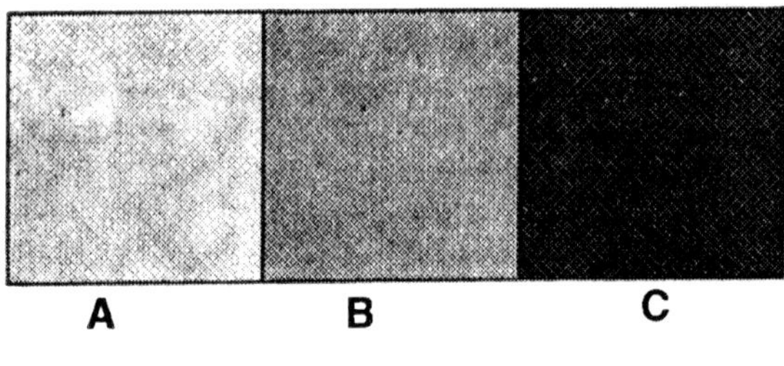

Figure 7.4

vacuum (e.g. Apiezon 'B') oil on it. The porcelain will indicate the amount of carbon evaporated-the oil spot remains clear and an easy visual estimate can be made. Figure 16.1 shows the visual densities required for different types of films. Under a high vacuum (10^{-3} torr) evaporate carbon by passing a high current (15 amp-20 amp at 30 v) through suitably ground carbon rods" held about 15 cms above the slide. The porcelain marker should be about as dark as Figure. Three ways of grinding carbon rods are shown in Figure 7.5 B; Figure 7.5 C is the method recommended by Balzers Ltd for platinum/ carbon replication.

The carbon rods for general work are ground to a wedge shape. The one flat and one thin dowel shape is tedious, but more suitable for obtaining an even evaporation rate of the carbon and keeping a measure of the amount of carbon lost. Only the thin portion will evaporate.

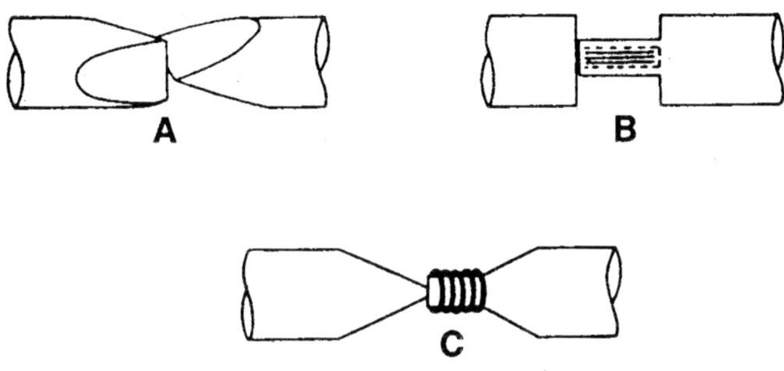

Figure 7.5

The time taken to evaporate this amount of carbon should be about ½–1 second. Carefully score the film into small squares and slowly lower the slide into a dish of distilled water. The surface should previously have been cleaned with fluflicss paper as described for collodion. The squares of carbon film will float on the surface of the water.

Take a grid in a pair of forceps and carefully immerse in the water. Bring it slowly up under a square of film and lift out of the water. Hold the grid in the horizontal position and, while still wet, remove the overhanging film with a sharp needle. Etched grids adhere better: dip the grids for a few seconds beforehand in concentrated hydrochloric acid or very dilute nitric acid and rinse thoroughly with water. Place the coated grids on filter paper to dry.

Method 2

Cleave a piece of mica and evaporate on to the clean surface a layer of carbon as described above and equivalent. Use a sharp needle to score the film very lightly. Immerse slowly in water and pick up the film as before.

Method 3

Coat the grids with collodion or Formvar as described above. Bedacryl 122x (available as a 40% solution in xylene) is also recommended for this purpose as it is easily dissolved away : dilute it to a 5 % solution with benzene. Place the coated grids on a clean glass slide and evaporate carbon on to them as before-less carbon should be used. Dissolve away the plastic film in a suitable solvent. Bedacryl is best dissolved away in acetone. Do not dissolve away the plastic if a double film is required.

Silica films are less generally used than carbon films as they are difficult to see and are quite brittle. Otherwise they are very suitable for high resolution work.

Perforated or 'holey' films

These are used mainly for the correction of astigmatism in high resolution work and as a support film for viruses. A virus film will sometimes dry unsupported across the small holes thus giving the maximum possible definition.

Method 1

Make a 2% w/v solution of Formvar in chloroform and add 0.5%-10% v/v water. The ratio of water to Formvar will affect the size of the hole, more water for large holes. Shake the mixture vigorously to obtain a milky suspension. Cast a film on water as for collodion and mount on grids. Allow the grids to dry and immerse for 30-40 seconds in acetone.

Method 2

Add a drop or two of glycerol to 100 ml of 0.2% w /v Formvar in chloroform. Shake to obtain a milky suspension. Coat a clean microscope slide with a film and dry. Place the slide in a steam bath for 1-2 minutes, remove, then float the film on water and pick up on grids. Allow them to dry. Larger holes may be obtained by adding more glycerol to the Formvar solution.

To make perforated carbon films coat holey-Formvar grids with carbon under vacuum and dissolve away the Formvar in chloroform.

REPLICAS

Surface detail of plant tissue, pollen grains or spores cannot be studied directly in the transmission electron microscope. Therefore a material that precisely makes a negative cast or replicates the surface is used. It must be coherent, non-crystalline and electron transparent. Moreover it must be easy to handle. For this purpose carbon is commonly used, sometimes silica and occasionally certain plastics.

Carbon replicas

Carbon evaporation is carried out in a vacuum coating unit; therefore fresh tissue must be treated quickly. Prolonged exposure to a vacuum causes gross distortion, cells lose water and rupture and surface waxes slowly volatilise. When the surface has been coated with carbon, the tissue is removed or etched away and the carbon mounted on grids for examination.

The carbon is removed by one of two methods. Either the tissue is etched away or the carbon is stuck to a plastic film, which can be peeled off the tissue. The latter method is more commonly used.

First take a sample piece of the cellulose adhesive tape used in your laboratory and dip it into acetone. If it curls up try other brands until you find one that does not curl.

Stick a piece of fresh tissue (if a leaf, not more than 2 sq cm) on to a clean glass slide using cellulose adhesive tape. Place under the bell jar of the evaporation unit. Next to it place, as for films, a piece of glazed white porcelain with a drop of high vacuum oil[20] on it as an indicator. Pump down rapidly to a vacuum of better than 10^{-3} torr using the rotary air pump; then expose for 1 minute to the diffusion pump. Evaporate 10-15 nm of carbon (equivalent to Figure 7.2 C) at a range of 15 cm. Remove quickly from the vacuum and proceed as in (a), (b) or (c).

(a) Flood the coated tissue with 2% w/v Formvar in chloroform. Drain vertically and allow to dry. The whole process, starting with the fresh tissue up to this point must be done in 5-6 minutes. Flood the specimen with 5% v/v Bedacryl 122x in benzene, drain and dry thoroughly (1015 minutes at room temperature). Do not allow the specimen to dry for longer than this or it will become brittle.

Lightly press a piece of the chosen cellulose adhesive tape onto the dry Bedacryl layer and gently pull off the slide. If the tissue remains attached, gently pull off with a pair of fine forceps.

Immerse the tape (carbon uppermost) in a bath of acetone, which will dissolve away the Bedacryl and the adhesive layer of the cellulose tape. Gently insert 200 mesh support grids between the Formvar and cellulose tape. Remove from the acetone, then carefully score around the perimeter of each grid with a sharp needle. Re-immerse in the acetone, then gently remove the coated grids. Wash in a fresh bath of acetone and allow to dry. Finally wash all the grids in two baths of chloroform to remove the Formvar. The carbon film can then be examined.

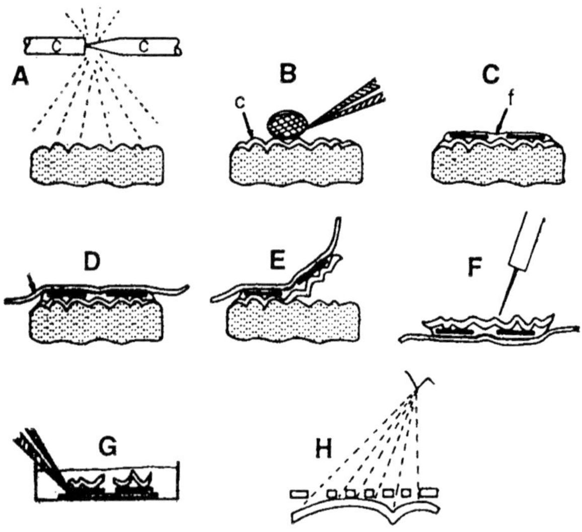

Figure 7.3

(b) Coat the leaf with carbon. Dip grids into 2% w/v Formvar in chloroform and gently place on to the freshly coated surface. Allow to dry. Quickly flood the surface with a little Formvar solution, drain and allow to dry. Lightly press a piece of cellulose adhesive tape onto the Formvar and carefully remove the tissue.

Score carefully around the grid perimeters with a sharp needle. Dip into a chloroform bath and remove the coated grids. Rewash in a fresh chloroform bath and allow to dry.

(c) Place the coated tissue in a bath of chromic acid. After ½ tol hour the tissue will separate from the carbon film. Remove carefully to a water bath, wash and mount on grids. This method is a destructive method in that the tissue cannot be reused. It has the advantage of cleaning the replica while etching away the tissue. Modifications of method (c) can also be used on spores, pollen grains or micro-organisms. Never attempt to make a carbon replica of a whole spore or pollen grain. The chromic acid etching will expand the material and burst the replica. Three-quarters embed the grains or spores in the surface of a soft layer of Formvar or Bedacryl and replicate the small region showing above the plastic. Then etch away the grains and the plastic.

SHADOWING

Carbon replicas or whole objects without shadowing give only limited information about the features of the surface relief and shadowing is often used to provide more information and to create a three-dimensional effect. The freshly coated grids are placed on a glass slide in an evaporation unit and, under a high vacuum, a heavy metal is evaporated at a known angle on to the grids. It

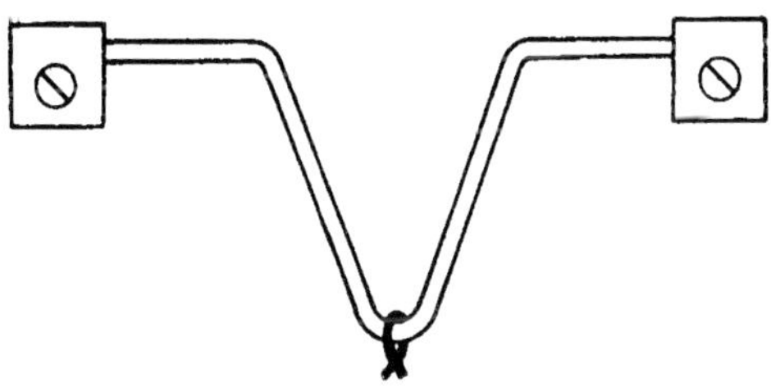

Figure 16.4

is useful to know the shadowing angle as rough calculations on the height and size of surface features may be made. The shadow lengths on photographic prints of a known magnification may be measured. Precise measurements are rarely possible since the exact position of a grid square relative to the metal source is rarely known. 45° is commonly used, 30° is often used on viruses and similar small objects. Several metals are used for shadowing; some are used for moderate resolution (× 1,000 – × 10,000) and others for high resolution (up to × 50,000). A gold/palladium combination is commonly used for low resolution work; tungsten, self-shadowed carbon and platinum/carbon are used for high resolution work.

Shadowing will leave regions in the 'lee' of objects which will have a low electron density. These regions will correspond to the darkest or shadowed regions on the plate. Thus the photographic negative is a true positive of the original surface and, for the greatest ease of interpretation, an intermediate negative should be made before it is printed.

Before setting up the bell jar for shadowing, place in position a 0.02 in. (0.5 mm) diameter filament of tungsten wire. Under vacuum heat the filament to red heat. This pre-heating will boil off contamination from the surface of the metal and will remove the stresses set up by bending it to a V-shape. It will also make the filament much more brittle.

Place the specimen coated grids on a clean glass slide in the coating unit. If you are shadowing a replica place the carbon side down (i.e. shadow through the grid bars). Around the tip of the tungsten filament wrap the shadowing metal wire approximately 5 mm of 0.005 in. (0.0125 mm) gauge wire. Arrange the grids so that they are for most specimens at an angle of 45° to the filament tip or a lower angle for small objects. Set them at a range of 7-8 cm for a single 5 mm loop or 8-9 cm for two metal loops (e.g. gold/palladium) from the filament. Evacuate the bell jar to 10^{-5} torr. Pass a current through the filament until it is a dull red colour. Slowly raise the current, watching the filament carefully through a smoked glass screen, until the metal forms a small blob on the filament tip. Continue raising the current slowly until the filament becomes bright white (10 amp at 30 v). Do not heat further as the filament may 'blow', depositing large pieces of tungsten on the replica. Switch off the current and allow the filament to cool. Remove the shadowed specimen for examination. This method is generally used for gold/palladium and platinum shadowing.

Tungsten shadowing produces sharp shadows and very fine grain and is ideal for high resolution work. The only disadvantage is the slow rate of evaporation.

Place the replica grids on a clean glass slide and place in the evaporation unit. Fine tungsten wire (0.005 in. or 0.0125 mm diameter) or a filament from the microscope electron gun is required. The tip of the filament is ground to half its diameter with very fine emery paper and mounted in the evaporation unit. This weak spot promotes evaporation from this point source. Evacuate the bell jar to 10^{-5} torr and slowly raise the filament current until the tungsten is incandescent white (2 amp-3 amp at 30 v). Note the ammeter current. As the evaporation proceeds, the current will fall-keep the current up to its initial value until the filament burns out. This takes 10-20 minutes.

Self-shadowed carbon replicas are not commonly used as the carbon tends to pile up in the 'lee' of surface features and destroys some of the original detail. Carbon does, however, produce very sharp shadows and is almost completely amorphous.

Platinum/carbon self-shadowed replicas have a very fine grain and should be used where the highest resolution is required. A shield with an aperture of 2-3 mm is often inserted between the source and grids to inhibit scattering and to improve the shadowing. Platinum/carbon amalgam rods can be made in the laboratory although the process is laborious and the rods are soft. It is also an expensive method, especially with rods purchased premade.

A simple method is to drill a 1 mm hole, 2-3 mm down the centre of a carbon rod. Turn the rod on a lathe to produce a 3 mm dowel with the hole down it. Insert into the hole 2 mm lengths of 125μm or 250 μm platinum wire, a total of 2 cm or 1 cm respectively. Alternatively, wrap the uncut piece of wire around a 2 mm dowel. Balzers supply a tool for cutting the dowel and winding the coil of wire. Place the surface to be replicated on a clean glass slide in the evaporation unit at the required angle and at 4-5 cm from it. Under a high vacuum (10^{-5} torr or better) pass a high current (20 amp-30 amp at 30 v) until the dowel has evaporated. Then evaporate a thin layer of carbon from vertically above to stabilise the replica.

Rotary or circular shadowing

This technique is only used for specific structures, e.g. for enlarging the apparent size of DNA molecules. A brushless electric motor of about 100 rpm has a small table mounted horizontally on the spindle. The specimen is placed on the table and while rotating, the shadowing material is evaporated on to it. Due to the structure of the motor (presence of insulating lacquer, etc.) pumping to a reasonable vacuum may take 2 hours or more.

PREPARATION FOR SCANNING ELECTRON MICROSCOPY

In the scanning microscope a beam of electrons is focussed by a condenser lens to a fine point on the specimen and made, by scanning coils, to scan or sweep over the specimen's surface in a 'raster' as in a television set. The electrons which hit or slightly penetrate the surface cause secondary electron emission. These secondary electrons, their number and energy governed by the nature of the surface from which they emerge and the energy which is applied to them, are collected by a scintillator/photomultiplier system. This collected signal is used to produce a scanned image on a long-persistent cathode-ray tube screen. The technique gives us a method of examining solid specimens directly and in many cases is the easiest to use if the surface is very rough or has other characteristics which make it impossible to examine in a transmission microscope. The resulting image has a marked three-dimensional character; the resolution lies conveniently midway between that of the light and electron microscopes and varies, depending on the accelerating voltage of the beam, from 0.2 μm at 3 kV to 20 nm at 20 kV.

All specimens in the scanning electron microscope build up a charge as the electrons strike them. Unless the surface is conducting this distorts the image. Non-conducting surfaces must be coated with conducting material to drain away the charge. Different types of specimens need different treatments.

Dry particles, e.g. pollen grains, spores, seeds, diatoms.

Coat the specimen stub from the scanning microscope with a thin layer of Formvar or a glue such as Durofix. Before the adhesive is dry scatter the specimens on to the sticky surface. Larger specimens such as seeds may be oriented to present various faces. Dry all specimens for 48 hours in a desiccating jar over silica gel or P_2O_5. Mount the stubs on the platform of the brushless motor in the vacuum bell jar. The motor should rotate at about 120 rpm. Under vacuum, deposit onto the rotating specimen a layer of carbon similar to that used for a normal thin film. Then also on to the rotating specimen deposit a layer of gold/palladium slightly thicker than that used for normal shadowing.

It is almost impossible to avoid handling the coating metal before hanging it on the tungsten filament. The contamination boiled off from the metal reduces the quality of the specimen. If possible, insert between the metal source and the specimen a shutter which can be removed as the metal approaches its melting point.

Wet specimens, e.g. bacteria in culture, unicellular algae.

Place a drop of the culture directly onto the specimen stub. The surface polymers of the specimens will stick them to the surface. Then dry and coat as above.

Delicate wet specimens.

Echlin has used accelerated freeze-drying to preserve details of specimens which would otherwise be damaged either by the vacuum or by drying. The organism on the specimen stub is plunged into either isopentane or Freon 13 cooled with liquid nitrogen or into liquid nitrogen at -158°C. The frozen specimens are then placed in an Edwards Tissue Dryer; the specimens are kept at –90°C and dried under a vacuum of 10^{-3} mm Hg. Depending on the size of the specimen drying will be complete in 3-4 hours. The specimens are then metal-coated in the usual way.

8

AUTORADIOGRAPHY

Autoradiography provides a way of tracing the path in the cell of an artificially supplied metabolite. This metabolite is 'labelled' by substituting a suitable radioactive isotope for one of the atoms of its structure. After fixation and sectioning of the tissue, a layer of photographic emulsion is placed on the section and the position of the isotope is shown by blackening of the emulsion.

Light microscope autoradiography is limited by the inability of the light microscope to resolve both the silver grains of the finest-grained emulsions and the fine structure of the specimen. The electron microscope, therefore, provides higher resolution of both the specimen and the emulsion layer. Further, the small aperture of the electron microscope permits both specimen and emulsion to be in focus together. This is impossible at high magnification in the light microscope, since the depth of field is much smaller.

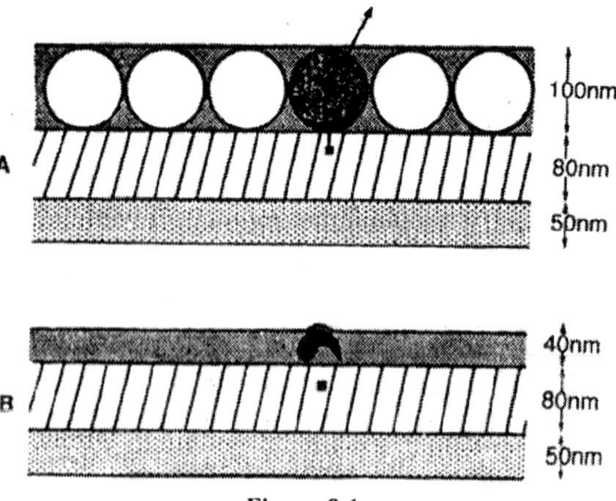

Figure 8.1

An autoradiograph preparation for the electron microscope normally consists of an ultrathin section, mounted on a grid in the normal way, coated with an emulsion consisting of a monolayer of silver halide crystals in a gelatine matrix. The crystals are large in relation to the thickness of the section, and it will be obvious that the resolution of the autoradiograph as a whole is poorer than that of the section alone in the electron microscope. (Although the developed grain, can be made smaller than the original crystal, its position is not related to the point at which the original crystal was hit by the beta particle.) The actual calculation of theoretical resolution is complex, but with the most commonly used emulsion the best attainable resolution is more or less the same as the original grain size, 0.1 μm.

The isotope almost universally used for electron microscope autoradiography is tritium (^3H). Only emitters of very low energy beta particles, which are highly ionizing and have little penetrating power, are suitable for high resolution work. This rules out most of the isotopes used in light microscope autoradiography. Much of the work done in this field on higher plants has used tritiated glucose[27] or other carbohydrates to follow the paths of wall-polysaccharide synthesis and deposition. Tritiated nucleotides[27] and amino acids have also been used, largely for bacterial work.

INCORPORATION OF LABEL

Consider the entire biochemistry of a metabolite when using it in autoradiography. For example, glucose is often used for studies of cell-wall deposition. However, it is im possible to tell the exact physiological fate of the supplied glucose. It can be incorporated in any of the cell-wall constituents, and also elsewhere throughout the cell. Roberts & Butt have pointed out that for many such investigations myo-inositol is more suitable. This is only incorporated in the matrix materials of the wall-pectins and hemicelluloses.

Use a compound of high specific activity, but not so high as to damage the tissue by radiation; electron microscope sections are so thin that otherwise insufficient label will be present to give a practicable exposure time. As a guide, activities commonly used for tritiated glucose are between 100 and 500 millicuries per millimole. Incubation times will depend on the tissue being studied. Because of the high cost of radio-isotopes in the activities

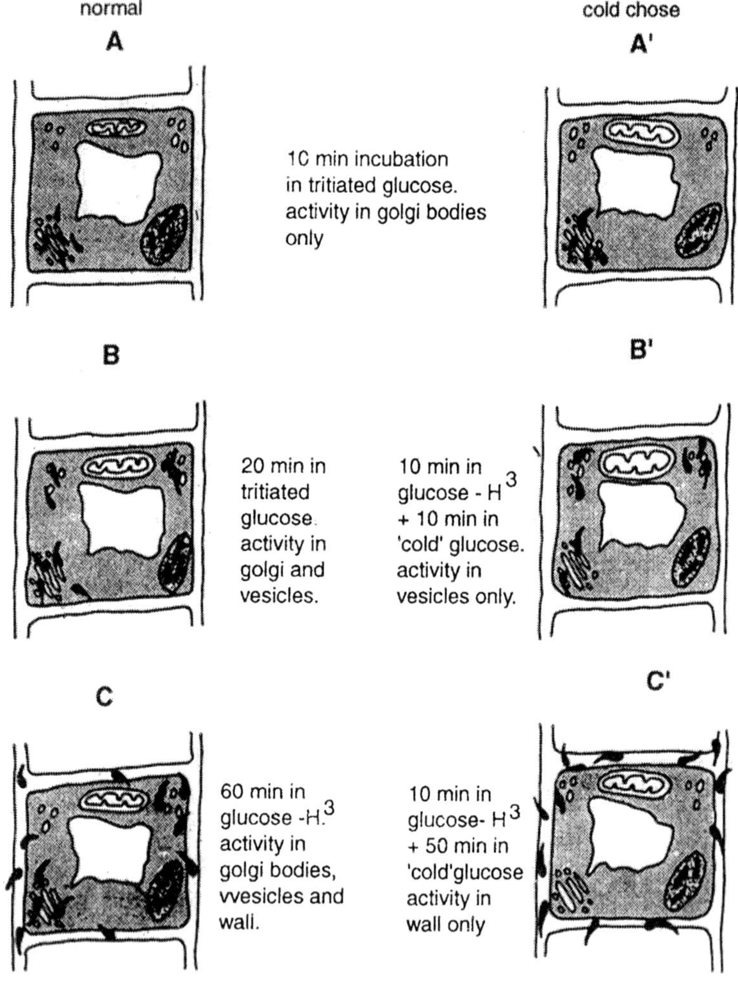

Figure 8.2

and concentrations required, it is normally only feasible to use small volumes of solutions. This is not a problem with algae and micro-organisms. In higher plants, tissue segments or detached seedling stems or roots must be used, with the attendant risk of morbid physiological reactions taking place. Small seedlings should be used where possible and they should be incubated with their root-tips in a drop of the solution.

The cold-chase technique is often useful to study a process taking place in a number of stages. Instead of incubating the specimens for varying lengths of time in the 'hot' (labelled) solutions, all the specimens are given the same incubation in the 'hot' solution and subsequent incubations of varying duration in an identical 'cold' (unlabelled) medium. Figure 18.2 shows the effect of the coldchase technique.

Check the amount of label taken up by the specimen by measuring the activity of the incubating solution before and after incubation in a scintillation counter. Similarly, check all fixing solutions and washes, at least on one run, to see if they are leaching out incorporated label from the specimen. Remove non-incorporated label before fixing with several washes of ice-cold distilled water. Fix, dehydrate and embed the specimens in the normal way.

Cut fairly thick (pale gold) sections, since these will contain more label than thinner sections, and therefore need a shorter exposure. Such sections will also have more contrast; this will help to counteract the loss of contrast due to the emulsion layer. The loss in resolution is unimportant, since it will still be far higher than that of the autoradiograph. Copper grids may react with the photographic processing solutions and, if used, should always be coated with a carbon or plastic film. Use stainless steel, palladium, titanium or gold grids where possible, all of which are available. These do not necessarily need a carbon or plastic film. The sections can be stained either before coating with emulsion, or after exposure and processing.

PREPARING AND APPLYING THE EMULSION

Use Ilford L4 for electron microscope autoradiographs. It is supplied as a shredded gel and has a shelf-life in the refrigerator of about 4 months. Melt before use and dilute to give a liquid emulsion which must be used immediately, as its shelf-life is nil.

L4 has a grain size of about 0.1 μm and is convenient to use since it readily forms a uniform monolayer. Emulsions of even finer grain are available (Kodak NTE, Gevaert NUC 307). However, these do not easily give uniform monolayers and suffer from severe loss of sensitivity during normal exposure conditions. Extremely complex procedures are needed to prepare and apply these emulsions for autoradiography, and they are not in general use.

The emulsion must be applied as a thin film containing a uniform single layer of silver halide crystals. Two methods of making this film are in common use, both devised by Caro & van Tubergen. The emulsion is applied to the section in these procedures as a gel rather than as a sol, making it easier to obtain uniform layers unaffected by irregularities in the section.

Loop Method

Carry out all the following operations in the darkroom.

Use a lime-green safelight, as for Bromide papers (e.g. Kodak Wratten OB).

1. Melt 10 g L4 emulsion in 20 ml distilled water in a 300 ml beaker for 15 minutes, using a water bath at 40-45°C. Stir the emulsion thoroughly without causing excessive frothing.

2. Transfer the beaker of emulsion to an ice-bath for 2-3 minutes.

3. Allow the emulsion to stand for 30 minutes at room temperature.

The emulsion should now be very viscous. Dip a loop of thin platinum, silver or copper wire, 4 cm in diameter, into the emulsion and withdraw it slowly. A thin liquid film should form across the loop. This film should very soon gel and cease to move within the loop. If it fails to gel, increase the time of cooling in the ice-bath. Arrange the grids on a glass microscope slide, attached if necessary at the extreme edge by double sided cellulose tape, such as Evo-Stick Twinstick Touch the gelled film to the surface of the slide and it will fall from the loop and adhere to the grids.

Agar Method

In this technique a smooth agar surface is prepared by casting on glass. The film is then cast on this surface, floated off, and transferred to the grids.

1. Make up a 2% agar solution in distilled water. Cast either by filling petri dishes to a depth of 0.5 cm, or, to obtain an even flatter surface, on plate glass. A convenient method is to make a rectangular tray to take a plate glass strip, as used for ultramicrotome knives. Fill the tray to give again a 0.5 cm depth of agar above the glass.

2. When the agar has cooled and set, it should be cut into 2 cm × 3 cm strips. Place these on microscope slides with the side cast on glass upward.

3. Warm the strips gently to remove surface moisture, either in a 37°C oven for a few minutes or with a lamp.

4. Flood the blocks with a 0.2% solution of Celloidini (Parlodion) in amyl acetate.' Drain them vertically and allow them to dry.

5. In the darkroom, melt and dissolve 10 g of L 4 in 40 ml of water, in a water bath at 40-45°C. Stir thoroughly and after 15 minutes remove from the water bath and allow to cool at room temperature.

6. Dip a 4 cm loop of platinum, silver or copper wire into the emulsion and withdraw it slowly (as in the previous method). Hold the loop vertically, to allow surplus emulsion to drain to the bottom. Absorb any surplus with filter paper. The emulsion will not gel-apply it as a sol to the coated agar blocks, by touching the film in the loop to the surface of the blocks. It dries very uniformly by the diffusion of water into the agar block through the collodion membrane. This avoids drying stains and gives a very uniform film.

7. Float off the emulsion-collodion layer, emulsion side upward, by lowering the block at an angle into a tray of water. It will float off very easily.

8. Place the sections on grids on a piece of wire mesh at the bottom of the tray. Pick up the film on to them by lifting the mesh under it.

Which of these methods to use depends very much on the type of work involved. The loop method has the advantage of simplicity. Further, it is capable of higher resolution, since it does not involve a collodion layer between the section and the emulsion. Some workers using pre-stained sections, however, prefer to use a film over the section, even with the loop method, to prevent chemical interactions between the stain and the emulsion or the processing solutions. Using the agar method the collodion membrane is unavoidable, so the maximum attainable resolution is lower. However, the layer of emulsion is more uniform, so this method is more suitable for quantitative work.

EXPOSURE AND PROCESSING

If the grids are on glass slides (Method 1) store them on these for exposure. Alternatively, use a grid box, modified by drilling a small hole in each compartment to allow processing fluids to run through when the autoradiographs are developed. During the exposure the grids must be kept in a completely light-tight box. A drying agent is also essential, to keep the air around the emulsion completely dry. Make an exposure box suitable for a grid box or a few glass slides, from an old plate carton, by fitting a cardboard false bottom with granules of silica gel under it. For a large number of glass slides, use a slide box with a bag of silica gel. Wrap it securely in a black polythene bag to make it light-tight.

There is a difference of opinion over the best temperature for storing the grids during exposure. Room temperature seems adequate for exposures of up to 3 or 4 months. At room temperature the emulsion has a higher sensitivity than at refrigerator temperatures. However, under some conditions a regression of the latent image, giving an apparent loss of sensitivity, can occur, and refrigerator storage can prevent this. In the exposure conditions of electron microscope autoradiographs, regression seems to be rarely as important as the loss of sensitivity at low temperatures, and refrigerator storage is only needed in hot climates.

The exposure time needed can only be determined by experiment. It depends on the amount of radioactive label present in the section, and this in turn depends on the activity of the original solution, the degree of incorporation and the thickness of the section. It may be only a few weeks, but in many cases will be two to four months.

Processing is carried out with the grids still attached to their glass slides or contained in their grid boxes. Pass the slides from dish to dish, agitating gently in each, exactly like prints. If you use a modified grid box, place it, minus its perspex cover, on a piece of filter paper in a Büchner funnel. Pour each solution into the funnel and after the required length of time draw it through into the flask with a filter pump in the usual way.

Table 4 below gives a routine processing schedule for L4 emulsion. The resolution obtainable with this is about 0.2 μm, Keep all the solutions at 20°C, and let grids that have been kept under refrigeration reach room temperature before development.

TABLE 18.1 : PROCESSING SCHEDULE FOR L 4 EMULSION

Developer	***Stop***	***Fix***	***Wash***
Kodak			
Ivlicrodol-X	10 seconds	Kodafix	*Slides:*
normal strength	1%	diluted 1: 4	running water
5 minutes	acetic	or other normal	5 minutes
or	acid	fixer.	distilled water
Ilford ID 19	or		1 minute
undiluted	distilled		*Boxes (in Büchner*
1 minute	water		funnel);
			distilled water
		1-5 minutes	10 minutes
			(several changes)

For the highest resolution (c. 0.1 μm) a physical developer can be used. This acts by first dissolving the silver halide, leaving only the latent image, composed of subelectron-microscopic silver grains. Silver ions from the solution are then precipitated on to these nuclei until visible grains are formed. The best formula is a solution of 0.1M sodium sulphites and 0.01M paraphenylenediamine. Dissolve the sodium sulphite first, in distilled water at 50°C, then add the paraphenylenediamine. Filter the solution before use. It must be used immediately, as it is unstable. The development is 1 minute at 20°C.

This developer gives very small, comma-shaped grains. The point of the comma is at the position of the original latent image. This defines the point at which the betaparticle hit the emulsion to within a limit of accuracy equal to the size of the original silver halide crystal (the latent image is not necessarily formed at the point where the crystal is hit). However, it is difficult to obtain consistent results with this developer. Both the sensitivity attained and the grain size tend to vary, although in both respects it is normally good. Use it only when the maximum possible resolution must be obtained.

Carry out post staining after fixation if the sections were not stained before exposure. Some workers have also used techniques for removing the gelatine from the developed emulsion, leaving only the silver grains. These may alter the positions of the silver grains and are not generally used. If required remove the gelatine and stain the section at the same time by using Karnovsky's alkaline lead stain at pH 12 for 30 minutes. It has been claimed that the gelatine can be removed in the electron microscope by irradiation with the beam at high intensity, after a stabilising irradiation with a low intensity beam. This procedure should be tried with great caution, however, if the section under examination is valuable !

AUTORADIOGRAPHY OF NUCLEIC ACIDS

Radioactive isotopes can be incorporated into cellular molecules. After the cell is labeled with radioactive molecules, it can be placed in contact with photographic film. Ionizing radiations are emitted during radioactive decay and silver ions in the photographic emulsion become reduced to metallic silver grains. The silver grains not only serve as a means of detecting radioactivity but, because of their number and distribution, provide information regarding the amount and cellular distribution of radioactive label. The process of producing this "picture" is therefore called *autoradiography* and the "picture" itself is called an *autoradiogram*.

Number of silver grains produced depends on the type of photographic emulsion and the kind of ionizing particles emitted from the cell. Alpha (a) particles produce straight, dense tracks a few micrometers in length. Gamma (y) rays produce long random tracks of grains and are useless for autoradiograms. Beta ((3) particles or electrons produce single grains or tracks of grains. High energy (3 particles (such as those produced by ^{32}P) may travel more than a millimeter before producing a grain. Low energy /3 particles (^{3}H and ^{14}C) produce silver grains within a few micrometers of the radioactive disintegration site and so provide very satisfactory resolution for autoradiography.

The site of synthesis of cellular molecules may be detected by feeding cells a radioactive precurser for a short period and then fixing the cells. During this *pulse labeling,* radioactivity is incorporated at the site of synthesis but does not have time to move from this site. The site of utilization of a particular molecule may be detected by *chase labeling*. Cells are exposed to a radioactive precursor, radioactivity is then washed or diluted away and the cells allowed to grow for a period of time. In this case, radioactivity is incorporated at the site of synthesis but then has time to move to a site of utilization in the cell.

In this exercise ^{3}H-thymidine and ^{3}H-uridine will be used to locate sites of synthesis and utilization of DNA and RNA, respectively. Uridine is the ribose-containing nucleoside of uracil and may be purchased with the tritium atom attached at various places on the purine ring.

Uridine is incorporated primarily into RNA but in *Tetrahymena* a small amount is converted into deoxycytidine (in other organisms it is even converted into thymidine) and incorporated into DNA. If appropriate controls are digested with RNase to remove RNA or extracted with hot acid to remove DNA and RNA the type of nucleic acid labeled with uridine can be confirmed. Thymidine, the deoxyribose nucleoside of thymine, can be purchased with the tritium label attached to the methyl group of thymine. Thymidine is specifically incorporated into DNA in *Tetrahymena.* Some organisms can remove the methyl group from thymine, and incorporate the uracil product into RNA. Even in this case RNA would not be labeled because the tritium label would be removed with the methyl group. Methyl labeled thymidine, therefore, serves as a very specific label for DNA. Controls will again be run with RNase digestion and acid extraction to check the specificity of labeling.

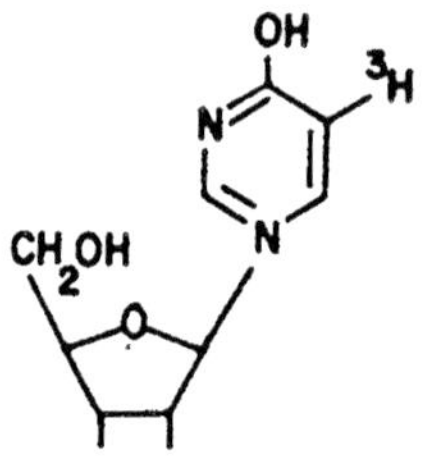

OH ^3H N N CH_2OH O

Uracil (5-^{3} H) riboside
(tritiated uridine)

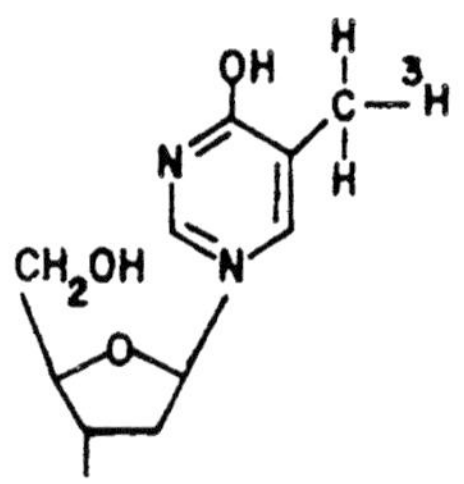

H OH C ^3H N H N CH_2OH O

Thymine(methyl-^{3}H)
deoxyriboside (tritiated thymidine)

PREPARATION OF MICROSCOPE SLIDES

Dip 18 (or more) clean slides in warm fresh gelatin-chrome alum to a depth of at least 2 in. Stack these vertically on the undipped end until they dry.

Using india ink (a rapidograph pen works fine) or wax pencil, mark a 1-cm diameter circle about 1 in. from the dipped end of the slide. Mark the undipped ends as follows:

1. 2 slides marked C (unlabeled controls)
2. 4 slides marked TP (^{3}H-Thymidine pulse label)
3. 4 slides marked TC (^{3}H-Thymidine chase)
4. 4 slides marked UP (^{3}H-uridine pulse)
5. 4 slides marked UC (^{3}H-uridine chase)

Rules for Safe Handling of Radioactive Material

1. All work with radioactive material must be done in a tray

lined with absorbent paper.

2. All glassware and equipment contacting radioactive material must be appropriately labeled and kept inside the tray. The only exception is that microscope slides of labeled *Tetrahymena* may be removed from the tray after the drop of labeled cells has been applied to the slide and allowed to dry.
3. Plastic gloves should be worn when handling radioactive material.
4. All waste solutions containing radioisotopes, all contaminated gloves, paper, etc., must be placed in appropriate liquid or dry *radioactive waste* containers.
5. To avoid spreading contamination, discard plastic gloves in the *Radioactive Waste* if they become contaminated. Put on clean gloves to complete the exercise.
6. Immediately report any spillage after first circling the area with a pencil, labeling that there has been a spill, and warning others to keep away.
7. Unauthorized personnel are not permitted in a radioisotope laboratory.
8. All authorized personnel must wear a radiation dosimeter (badge) while in the radioisotope laboratory.
9. Wash hands thoroughly with soap and water as soon as you have finished handling radioisotopes and removed the plastic gloves.
10. All pipetting must be done with a pipettor in this laboratory.

PREPARATION OF EXPERIMENTAL SLIDES

Refer to Figure 8.3 for a flow diagram of the experimental procedure.

Growth of experimental culture

Inoculate a 30-ml milk-dilution bottle culture of *Tetrahymena*. Grow this at room temperature for 24 hrs.

The culture should then contain enough cells to prepare several hundred microscope slides but still be in log phase.

Centrifuge the culture for 2 min at 2000 X g and discard the culture medium. Resuspend the cells in 3 ml buffer and place 1 ml of this cell suspension into each of 3 small centrifuge tubes. Proceed simultaneously with the following three aspects of the Exercise.

Unlabeled controls **Mark one centrifuge tube C**

Let these cells stand for 1 hr then add 1 drop concentrated formalin. Spot small drops of cell suspension in the circles on the 2 unlabeled control C slides and allow to air dry. During the hour wait, proceed with the labeling.

^{3}H-uridine pulse labeling

Be sure that all of the following steps are carried out in the paper-lined tray.

Mark a 2nd centrifuge tube U. Add 50 pC ^{3}H-uridine to this cell suspension. Let the cells incubate for 5 min at room temperature.

Put 1 drop of formalin in a small centrifuge tube and label this UP. When the cells have incubated for 5 min, transfer half the cell suspension to this UP tube using a clean Pasteur pipette. The formalin will kill the cells and stop uridine incorporation at this time.

Immediately centrifuge the remaining live cell suspension and remove the radioactive supernatant with a clean Pasteur pipette. Place this supernatant in the liquid radioactive waste container.

Resuspend the cells in buffer and recentrifuge. Again pipette the supernatant into the radioactive waste. Resuspend these chase (UC) cells in 1/2 ml fresh culture medium and let them incubate.

The formalin fixed UP cells can now be washed once by centrifugation. Place supernatant in the liquid radioactive waste container and resuspend the cells in buffer. Spot cells in the circles on all UP labeled slides and air dry.

^{3}H-thymidine pulse labeling

Mark the last centrifuge tube T. Add 50 pC ^{3}H-thymidine to this tube. (This can be done during the 5 min uridine pulse labeling.)

Let these cells incorporate thymidine for 15 min, then kill half the cells by transferring an aliquot to a centrifuge tube containing 1 drop formalin. This tube of killed cells is labeled TP and can be spotted at any convenient time.

Immediately centrifuge the remaining live cells. Pipette the supernatant into the liquid radioactive waste container.

Resuspend cells in buffer and recentrifuge. Place supernatant in radioactive waste. Resuspend these chase (TC) cells in 1 /2 ml fresh culture medium and let them incubate.

Wash TP cells by centrifugation. Spot on TP labeled slides and let air dry.

PREPARATION OF CONTROL AND CHASE SLIDES

A total of one hour after labeling was begun, kill all cells by the addition of 1 drop formalin. Wash the cells once by centrifugation and resuspend in buffer. Spot the chase groups of cells. Be sure to use separate clean Pasteur pipettes for unlabeled controls C, uridine chase UC, and thymidine chase TC cells. Let all slides air dry. Slides should be removed from the paper-lined tray at this time, and can be stored until later.

Fixation, digestion, and extraction

a.	Fix all slides (C,UP,UC, TP, TC) in acid-alcohol	5 min
b.	Rinse gently in distilled water	2 min
c.	Digest 1 slide from each labeled group (UP, UC, TP, TC, *not C) in* buffered 0.2% RNase at 37°c. Label these slides RNase.	1 hr minimum
	Extract a second slide in either 2N HCl or 5 % perchloric acid at 80°C. Label slides appropriately.	1 hr minimum
d.	Rinse all slides in several changes of distilled water	2 min

PREPARATION OF AUTORADIOGRAMS

All of the following steps must be carried out in complete darkness or with no more than a 15-W bulb covered with a Wratten series 2 safelight filter kept at a minimum distance of 3 ft from the photographic emulsion.

1. Warm some Kodak NTB or NTB_3 bulk emulsion in a water bath at 40°C for 30 min.
2. Pour some emulsion into a slide dipping vial (i.e., the smallest volume container that will allow a microscope slide to be dipped into it, to a depth of 2 in.; a 30-ml beaker works fine). Clamp the vial stationary in the 40°C bath. Return the remainder of the bulk emulsion to a light-tight box.

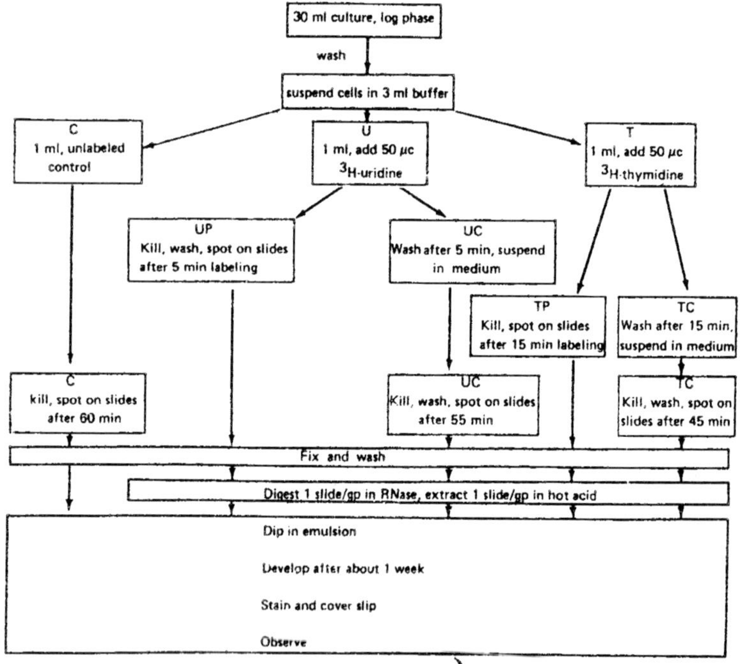

Figure 8.3 : ***Flow Diagram for autoradiography of nucleic acids.***

3. Hold a slide by the labeled end and dip the other end into

the warm emulsion until the circled area of dried cells is completely immersed. (Two slides may be held back to back and dipped simultaneously to conserve time and emulsion.)

4. Stack the dipped slides in a slotted board, label end down and continue dipping until all experimental slides are dipped and stacked.
5. When the emulsion has air dried (about 30 min) place the slides in a light-tight slide box or in glass slide trays that can be placed inside light-tight cans. The container should contain a drying agent such as drierite and be sealed with photographic tape.
6. Store the slide container in a refrigerator (approximately 4°C *not* freezing) for 1 week.

Development of Autoradiograms

The first three steps must be carried out in the dark as in 1 V.

1. In the dark, develop slides in Dektol or D-19 at 18°C. — 2 min
2. Rinse slides gently in distilled water, 18°C.
3. Fix in Kodak acid fixer at 18°C. — 8 min
4. Lights can be turned on and slides rinsed in running tap water. — 30 min
5. Stain slides in 0.1% fast green at pH 2.5. — 1-3 min
6. Rinse in distilled water, — 1 min
7. Rinse in fresh distilled water. 1 min
8. 70% alcohol — 5 min
9. 95% alcohol — 5 min
10. 100% alcohol — 5 min
11. Xylene — 5 min
12. Mount in Permount as described in Exercise 5.

ANALYSIS

Labeling Patterns

Tritium autoradiograms of whole cells have sufficient resolution to determine three cellular sites of labeling. You can see if the macronucleus or micronucleus or cytoplasm is labeled. Cytoplasmic labeling shows as a random array of black silver grains over cytoplasm with a grain density at least 2 times that of background on the same slide. Macronuclear labeling appears as a large solid area of silver grains about 15 m in diameter. Micronuclear labeling is characterized by small areas of very intense labeling. Frequently, over a dozen silver grains are crowded in each of 2, approximately 3 m diameter areas.

Raw Data

Grain counting can be done to determine if labeling has occurred or to determine amount of labeling. In this exercise labeling should be obvious and it is only necessary to count the number of cells observed in each label category. In Table 8.2 record the number of cells observed in each label category for each experimental treat-ment. Unless gross artifacts are present, cells should be found only in the label categories.

Treated Data

From the total number of cells counted for each experimental treatment, calculate the percent of cells in each label category (include percentages in Table 8.2). Plot these percentages (y-axis) as a histogram for the label categories observed (x-axis). Plot different labeling treatments and extractions as separate histograms under one another. Thus, cellular label categories can be compared from left to right and the effect of an experimental treatment can be compared from top to bottom (as in Table 8.2).

Interpretation of Data

Does *Tetrahymena* incorporate detectable amounts of ^{3}H-uridine into a precipitable macromolecule? What is the major cellular site of incorporation (syntheis of the macromolecule)? By comparing UP, UP + RNase and UP + hot acid data, can you conclude that uridine is incorporated into RNA? Into DNA? Do

TABLE 8.2 : NUMBER AND PERCENT OF *TETRAHYMENA* CELLS LABELED IN DIFFERENT CELLULAR COMPARTMENTS AFTER ^{3}H-URIDINE OR ^{3}H-THYMIDINE INCORPORATION

Cell Treatment		Label Category							
		Cyt	Ma	Mi	Cyt + Ma	Cyt + Mi	Ma + Mi	Cyt+Ma+Mi	Total
Unlabeled control	No.								
	%								
UP	No.								
	%								
UP + RNase	No.								
	%								
UP + hot acid	No.								
	%								
UC	No.								
	%								
UC + RNase	No.								
	%								
UC + hot acid	No.								
	%								
TP	No.								
	%								
TP + RNase	No.								
	%								
TP + hot acid	No.								
	%								
TC	No.								
	%								
TC + Rnase	No.								
	%								
TC + hot acid	No.								
	%								

where Cyt = cytoplasmic lable, Ma = macronuclear label and Mi = micronuclear label

different cells incorporate uridine at different rates? By comparing pulse with chase data can you conclude that RNA migrates from its site of synthesis?

Does *Tetrahymena* incorporate detectable amounts of ^{3}H-thymidine into a precipitable macromolecule? What cellular sites of incorporation can you detect? From the extraction data can you conclude that thymidine is incorporated only into DNa? Do the cellular sites in which you detected incorporation, always show incorporation of thymidine? Does your data show any evidence of migration of DNA from its sites of syntheis (changes in frequency in lable categories from pulse to chase)? If your data clearly shows cytoplasmic incorporation of thymidine, does your resolution in autoradiograms allow you to determine the specific site of cytoplasmic DNA synthesis?

EFFECTS OF ULTRAVIOLET RADIATION ON CELLS

Much can be learned about the effect of an environmental agent on an organism by simply observing changes in behavior and morphology. The observer will note that ultraviolet (UV) radiation produces several gross cellular changes. Longer exposure increases the severity of these changes until death results.

The mercury vapor germicidal lamp, which is used as the ultraviolet source, produces some visible light but about 85% of its energy is emitted at 253.7 nm. In order to produce a biological effect, radiant energy must be absorbed by the irradiated material. Both nucleic acids and proteins (see Exercise 3) absorb light of this wavelength.

Absorption of UV light by DNA results in, among other things, the formation of thymine dimers which block replication of DNA until repair can be accomplished. If repair is not possible or is delayed the cell can be "sterilized" or—killed. This mechanism of UV damage will not be observed for at least a generation. Under the conditions of this exercise, one half of the *Tetrahymena* in a sample should be "dead" in about 5 min. Since *Tetrahymena* has approximately a 3-hour generation time, DNA damage can not be the cause of death. Such an immediate mechanism of killing must result from damage to indispensible cell enzymes or structures.

In many instances, electromagnetic radiation affects a chemical or biological system according to the BunsenRoscoe law:

Dose which produces a biological effect

= light intensity × duration of exposure

In this exercise, the biological effect to be measured is the "death" of one half of the *Tetrahymena* in a sample. In this case:

$$LD_{50} = I \times t_{50}$$

where LD_{50} is the lethal dose for 50% of the organisms, I is the intensity of the radiation and t_{50} is the time to kill 50% of the organisms. This law can be tested by determining the dose of ultraviolet radiation that is necessary to "kill" the organisms at different intensities.

METHODS

Demonstration of Procedure

Place an uncovered Petri dish containing a suspension of *Tetrahymena* directly beneath the ultraviolet light. At 30-sec intervals, stir the dish thoroughly and remove a small aliquot containing between 10 and 25 organisms. Place this small drop of *Tetrahymena* in the well of a depression slide. Count the total number of organisms and determine the percent "killed." Exposure to ultraviolet causes progressive changes in the behavior of *Tetrahymena*. Each individual will have to determine which of these changes will be called "death."

One method of observing *Tetrahymena* involves a form of dark field microscopy. The sample may be illuminated from below the stage by light entering the sample as nearly parallel to the stage as possible. Alternatively, the center of the condenser lens may be blacked out with a paper disk allowing only side illumination. Both of these methods result in brightly refractile *Tetrahymena* against a dark background.

Warning: Do not look even briefly at the germicidal lamps. The radiation can kill the conjuctiva cells of your eyes and cause pain and several days of bindness.

Determination of LD_{50}

Expose several samples of *Tetrahymena* to ultraviolet light, at one measured intensity. Determine an average t_{50}. Due to the variability of biological material, a single determination is not sufficient. An average of several samples is needed for accuracy.

Nature of the Lethal Mechanism

Run a series of *Tetrahymena* samples to determine if light really is essential to the killing mechanism and if UV is the only light involved in "killing." Since ordinary glass is impermeable to UV a Petri dish cover could be used as a UV filter.

Also test your criterion for judging "death." Take samples in which half the cells have been "killed" by UV and store some in the dark and some under visible light for a period of time (perhaps 30 min) to determine if recovery or photoreactivation occurs.

Proof of the Bunsen-Roscoe Law

Measure the intensity of the ultraviolet radiation using an ultraviolet sensitive photovoltaic cell and an ammeter. Use the wire screens to reduce light and provide the desired range in light intensity. Determine the t_{50} at several intensities. Measure the length and width of several typical cells.

ANALYSIS

Determination of LD_{50}

Describe the progressive behavioral and morphological changes observed in the *Tetrahymena* cells with UV irradiation. How did you determine "death"? What damage appeared to cause death of the cells? What was the average UV dose which killed 50% of the population? What was your variability in estimating the LD_{50} among the different samples of cells tested? Did the LD_{50} vary from day to day or among different cultures of cells on the same day? Why?

Nature of the Lethal Mechanism

Can you be certain that UV light, not visible light, heat or desiccation is responsible for the observed "killing"? What is photoreactivation? What type of macromolecule is repaired during photoreactivation? Do the *Tetrahymena* cells show any repair or recovery in the dark or in visible light? Based upon all your experimental evidence to this point, what is the most probable mechanism of "killing"?

Proof of the Bunsen-Roscoe Law

Calculate the LD_{50} for each intensity used. Plot LD50 (y-axis) against intensity (x-axis). Is the LD_{50} constant within the confidence limits of the experiment? Is LD_{50} constant with changes of intensity? Why?

If a calibrated photovoltaic cell is available, convert doses to ergs/mm^2. Next calculate the LD_{50} in quanta/ mm^2, using the energy (in ergs) of a quantum of light. Assume that all radiation in this exercise is at 253.7 nm. The energy of a quantum (0) is given by:

$$Q = \frac{hc}{\lambda}$$

where *h is* Planck's constant, 6.624×10^{-27} erg-secs, c is the speed of light, 3×10^{10} cm per sec, and λ is the wavelength of light in cm.

From the measurements of *Tetrahymena* estimate the cross-sectional area of a cell and the number of quanta penetrating each cell. From the number of quanta needed for "killing" can you decide on a cellular site of UV damage?

9

THE INSTRUMENT AND PHOTOGRAPHY

RESOLUTION

As in all types of microscopy, the resolution of an electron microscope determines its scope. Resolution is defined as the distance between two points which can just be seen as separate objects; the naked eye has a limit of resolution of about 0.1 mm, the light microscope a maximum theoretical and practical resolution of 0.2 μm and the ultraviolet light microscope a resolution of 0.1 μm. The maximum resolution of the electron microscope has not yet been reached, although it far exceeds that of the light microscope. At present its limit is about 0.3 nm.

In the light microscope as much or more information may be gathered by viewing the specimen as by studying micrographs of the same image. The depth of field of a good light microscope is considerably less than the thickness of an average section. This means that the maximum information from a specimen can only be gathered by focussing up and down through the depth of the object. The multiple images that the human brain then integrates cannot be recorded on a photographic plate. The image seen in an electron microscope is of very poor quality in comparison and photography is an essential part of electron microscopy. Moreover the depth of field of even the best electron microscopes is far greater than the thickness of a section.

Unlike the specimen in a light microscope, a section in an electron microscope is subject to deterioration, normally a slow but continuous process. As a general rule a satisfactory image in an electron microscope should be photographed before being studied, not after.

THE SCREEN IMAGE

Several factors control the quality of the image on the screen.

1. The viewing screen itself.

2. The stability of the circuits of the beam and lens systems and the column alignment

3. Specimen preparation

4. The vacuum in the column

5. The cleanness of the column

6. The objective aperture size

7. The accelerating voltage (kV) applied to the electrons.

The viewing screen

The viewing screen serves to delineate the area that can be photographed, and is used for aligning the microscope, correcting its astigmatism and focussing the image. Fortunately the depth of focus of even the best electron microscopes is measured in hundreds of metres so that the position of the screen in relation to the photographic plate is unimportant. The screen is not used in the actual photography and the poor image quality it produces is of no concern. The grain size of a screen is far greater than the grain size of a good photographic emulsion, the image contrast is poor and the light intensity low.

The stability of the microscope

The general maintenance of an electron microscope is beyond the scope of this book and it is assumed that the circuits are adjusted and the column aligned to give peak performance. The filament, when new, may move slightly at first, but this will only affect the centre of illumination. However, during the last part of a filament's life the filament current may oscillate slightly as the tungsten evaporates and finally parts. Do not take photographs if it is suspected that the filament is reaching the end of its life.

Specimen preparation

Specimen preparation and section cutting have already been discussed. As a general guide, for the very highest resolution, use only grey sections; for instrument magnifications up to x 20,000, silver sections are adequate and, for autoradiography, use gold sections since the primary consideration is the concentration of radioactivity and not the resolution. For high resolution work do not keep any electron microscope preparation longer than necessary. In particular negatively stained preparations deteriorate very rapidly. Even tissue embedded in resin deteriorates slowly.

The vacuum in the column

The vacuum in the column is critical. Although most instruments can be used at pressures just below 10^{-3} torr, there is a marked improvement in the contrast of the specimen as the vacuum improves. For the highest resolution do not take pictures until the vacuum has fallen below 10^{-4} torr.

Cleanness

The cleanness of the column affects picture quality in two ways. Firstly, dirt on the specimen or the instrument can cause apparent image movement by allowing an irregular charge build-up and leakage. Jumpiness of the image or apparent movements of the lens settings are much more likely to be caused by dirt than by electronic faults. Secondly, in instruments without a cooled specimen stage, dirt, consisting of burnt diffusion pump oil, vacuum rubber and rotary pump oil vapour, falls on to the specimen. In instruments without such cooling the image on the screen should be photographed within thirty seconds or abandoned.

Aperture size

Aperture cleaning is beyond the scope of this book, but the correct use of apertures is critical in obtaining the very best image. Follow the manufacturer's instructions in the choice of size for the condenser aperture. 250-500 μm hole diameter are the sizes commonly recommended. The choice of the objective aperture is determined by the specimen. The smaller the hole, the higher the contrast will be on the negative. A rough guide to their use is given in Table 5 below.

TABLE 9.1

Diameter of hole	***Specimen***
75 μm	Metal-shadowed replicas or shadowed whole objects, e.g. microfibrils, diatoms, algal scales, flagella.
50 μm	Routine well-stained sections; some unshadowed whole objects.
30 μm	High magnifications of very thinsections;negatively-stained preparations; autoradiographs and histochemical staining of most kinds.

If an aperture size smaller than 75 μm is used on high contrast objects in a modern electron microscope, it may be difficult to produce an acceptable print on the normal range of photographic papers. The 50 μm size is most commonly used and many microscopes without cooling stages cannot make effective use of 30 μm apertures because they contaminate too rapidly.

One way of overcoming the contamination problem may be to use evaporated apertures. These are available in the United States, but have not so far been widely used in this country. They are made from evaporated gold; they are about 0.5 μm thick and are said to have a long life and to be virtually free from contamination problems. They can be obtained in a wide range of sizes.

The accelerating voltage

The voltage used to accelerate the electrons along the beam path will directly affect the quality of the image seen on the screen. The higher the velocity of the electrons, the greater the thickness of specimen that can be penetrated, but the lower the screen contrast will be. As a general guide use accelerating voltages around 40-50 kV for normal biological specimens. It may be necessary to go up to 75 kV for autoradiographs or other thick specimens. No biological specimen should need voltages in excess of 75 kV.

THE PHOTOGRAPHIC IMAGE

A number of other factors may not markedly affect or be relevant to the quality of the screen image, but will seriously affect the print quality or the information that can be derived from the print.

1. Astigmatism
2. Focus and calibration of the instrument
3. Type of emulsion
4. Exposure time
5. Development, storage and cataloguing
6. Printing

Astigmatism

The correction of astigmatism both at the condenser and objective levels varies so much from instrument to instrument that little use would be served by giving a description here. Only a few general points can be made. The way to make holey-Formvar films is given on page 60. The smaller the objective aperture the more difficult the correction will be, and the shorter time any such correction will last. Evaporated thin film apertures although more difficult to make will last longer free from contamination than the conventional platinum or molybdenum disc. Beginners are advised to start with large objective apertures and work downwards. Electron microscopes with cooled-stages will stay corrected longer than those without. Apertures in instruments without cooled stages become fouled more rapidly with the uncollected contamination in the column. As a rough guide, correct the astigmatism at twice the maximum magnification you intend to use for the negatives, e.g. correction at × 120,000 will allow pictures to be taken up to × 60,000.

Focussing the image

A little must now be said on the actual operation of the instrument. One set of controls in almost continual use will be that governing the objective lens current-the focussing. It is essential to master this before good photographic images can be obtained.

Most electron microscopes provide a binocular viewing microscope above the viewing screen. It has a low power, × 7-- × 10, and is used solely as an aid in focussing and astigmatism correction.

For the newcomer to electron microscopy, focussing may be initially difficult and frustrating, but certain aids help matters. Small holes in very thin sections, pieces of inorganic dirt on the section or dense particles purposely added to the section, such as latex suspension or colloidal gold may be used as focussing aids. At very low magnification, focussing is often made easier by removing the objective aperture from the beam path and adjusting the objective lens current so that the image appears at a minimum contrast. The aperture must then be replaced before photographing the image.

At moderately high magnifications, initially one can focus on a portion of the specimen having some piece of dense matter or hole; the actual area to be photographed is immediately adjacent, but out of the beam path. This allows a relatively uncontaminated area finally to be photographed. For high resolution work this procedure is not recommended and a through-focal series should be taken.

Calibration

Periodically the magnification of the instrument must be checked. This is most easily achieved by using at low range a replica of a diffraction grating and at high ranges particles of a specific dimension.

Polystyrene latex particles, available as a suspension in a range of diameters may be used to check low magnifications. The particles are dense and easily photographed.

From measurements made however it was found that numbers of the particles were ± 10 % of their stated size and therefore not strictly accurate.

Cross-ruled diffraction gratings, or carbon replicas made from them, are available. The usual sizes are 590, 1180 and 2160 lines per mm (approximately 15,000, 30,000 and 55,000 lines per inch).

To calibrate an existing diffraction replica on an E/M grid, measure the width of a number of grid squares in a light microscope with a micrometer eyepiece and calibrated objective. Then in the electron microscope count the number of diffraction lines spanning one grid square and calculate the exact spacing. The 2160 lines per mm grid is only effective up to × 30,000 magnifications, above this range other objects must be used.

A suspension of Tobacco Mosaic Virus (TMV) particles is often used to calibrate ranges × 30,000 and above. TMV has a diameter of exactly 17 nm. Dry a drop of the suspension down on a support film and photograph ten or more rods packed closely side by side to ensure greater accuracy. Ferritin, which can also be used, has an iron core which is 5.5 nm in diameter.

Platinum phthalocyanine crystals with a lattice spacing of 1.25 nm diameter can be used for very high magnification calibration.

The particles are photographed at each stage in the magnification range of the instrument, the final negative being enlarged at a known magnification, e.g. × 5, and measurements made directly from the print.

PHOTOGRAPHY

Most modern electron microscopes have built-in cameras capable of taking either plates or film. Older models use only glass plates.

Film has a few advantages over plates; more exposures may be taken on one load, it is cheaper and takes up less storage space. It is, however, easily scratched, requires a longer time to pump down in the microscope, is very brittle under vacuum and individual exposures may be more difficult to label. Plates generally have a harder emulsion, are not so easily damaged and pump down more quickly in the microscope. They are dimensionally stable and should always be used if very accurate measurements are to be made from the image. They can easily be labelled and viewed individually. Plates and films are often prepumped or desiccated in an attachment to the microscope rotary pump.

Type of emulsion

Emulsions differ little in their speed in response to electrons; emulsions varying by a factor of 10^5 in speed to light differ by less than 10 to electrons. Therefore fast panchromatic emulsions simply complicate dark-room procedure with little, if any, gain in photographic quality and a possible increase in the graininess of the final image. Therefore any fine grain, non-colour sensitive emulsion is suitable. Examples of films and plates used are: for the Phillips E.M. 100, 200 and 300 and other microscopes using 35 mm cameras, Ilford 35 mm unperforated Fine Grain Safety Positive Film or Agfa Scientia 22D50 (35 mm) film. Electron microscopes using 70 mm cameras such as the G.E.C.-A.E.I. E.M.6B can use Agfa Scientia 19D50P or Ilford 70 mm recording film 5B11.

Plates suitable for many microscopes, including the Siemens Elmiskop series and the G.E.C.-A.E.I. E.M.6, 6B and 8 series are Ilford Special Lantern Contrasty and Kodak Electron Image plates.

Exposure time

Film or plate exposures vary between 1 and 4 seconds under normal conditions of screen illumination. When photographic exposures are made in the microscope it is essential that an even illumination of the image is achieved and that uniformly repeatable exposures are made. Most modern microscopes have a built-in exposure meter, calibrated for the particular photographic emulsion in use. Exposures are generally kept short, 1 to 4 seconds is normal practice. Periods shorter than 2 seconds are difficult to time accurately; also flap shutters on some microscopes would produce an unevenly illuminated plate at this exposure level. Above four seconds, resolution might be impaired by specimen movement, 0.1 nm drift per sec. being sufficient to destroy high resolution work.

To obtain consistent uniform density plates or film, several points should be remembered. Even illumination of the image is required; exposure by the flap-type shutter must be quick and positive; development must be under uniform time/temperature conditions. Intermittent agitation in the developer is essential.

Development

Normally, development is kept to a reasonably short time, 3 to 4 minutes being suitable. Development of the film or plate may be carried out in most contrasty developers. Ilford PQ Universal developer, diluted l in 10, Ilford 1D-2 diluted 1 in 5, Kodak DK-50 full strength or Kodak D-163 diluted 1 in 4 are all suitable for plates and film. 1D-2 produces a softer image and is suitable for contrasty subjects such as surface replicas. The film or plate is immersed in the developer and intermittently agitated for 3 to 4 minutes, depending on the degree of density required. It is then rinsed in tap water or a 2% acetic acid stop bath and fixed for 6 to 10 minutes in a hardener/fix bath. Kodak Kodafix, diluted 1 in 4 or Ilford Ilfofix, used full strength, is suitable; 6 minutes at 20°C is necessary for complete fixation. Use intermittent agitation. The film or plates are washed in cold running tap water for half an hour; plates are then allowed to drain and dry in racks; film is hung up -both must be dried in dust-free places and drying cabinets are recommended.

Storage and cataloguing

Processed films or plates are best stored in their original cartons. Plates should be stored on their sides. Negative envelopes

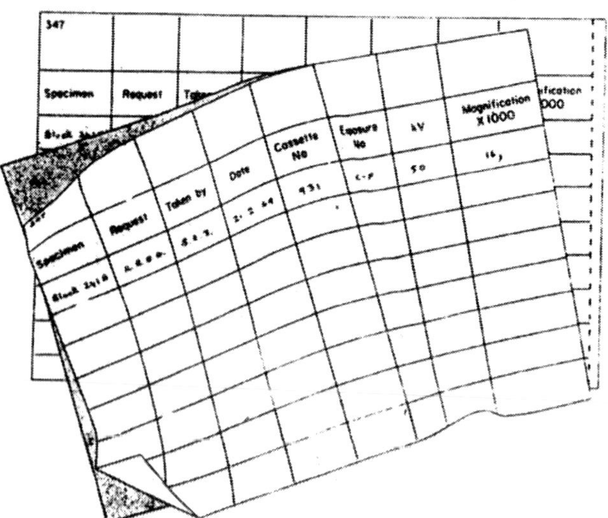

for plates in clear cellophane are available and they prevent scratching. Film and plates may be labelled before or after processing using an HB pencil or ballpoint pen on an unused portion of the emulsion.

A logbook for each instrument must be kept. A specimen entry from a specially printed logbook is shown below. Each page is duplicated. The carbon copy remains permanently in the book. The top sheet is perforated and may be removed by outside workers using or having work done on the microscope. The ruling is continued onto the right hand side of the page so that the back of the carbon copy of the next page is available for entries such as notes on the performance of the instrument or further details of the specimens examined. Such specially printed sequence numbered logbooks are expensive. Personal logbooks such as the Collins Cathedral series 69/5B1 may also be kept and are cheap to buy.

Printing

Printing is the last, but still an important part of processing. A high quality calibrated enlarger is required. Ideally it should be capable of reducing negatives to 2 in. by 2 in., the standard size of slides for talks and lectures. Also it must be capable of enlarging up to about × 10 or × 12 without any trace of detail loss at the edge of the negative.

As most negatives will be of reasonably low contrast, a 'hard' paper will generally be required for printing, e.g. Kodak or Ilford paper 3 or the equivalent Agfa paper N. Making the correct exposure will be a matter of practice. Usually a test strip is convenient to make and wastes only one sheet of paper if all negatives in a particular batch are of the same density.

Development of the paper in either Ilford PQ Universal diluted 1 in 12, Kodak DK 19 undiluted or D-163 diluted 1 in 4 are suitable. Prints must then be rinsed in a stop bath or tap water and fixed in a fixer/hardener bath. Again Kodak Kodafix diluted 1 in 7 or Ilford Ilfofix diluted 1 in 3 are suitable, fixation taking 6 to 7 minutes. The prints must then be washed for 50-60 minutes in cold running tap water. After washing, prints should be soaked for 5 minutes in distilled water containing Kodak PhotoFlo 200 or Ilford Bango glazing solution. They both prevent smears appearing on the prints and give a good even glaze.

Glazing may be carried out by several means. A flatbed type glazer is quite adequate for a small number of prints, but for heavy work, a rotary glazer, e.g. the Kodak Rapid Glazing machine or the Johnson Super-Gloss dryer is ideal. Thermostatically controlled, a rotating chromiumsteel drum is kept at a constant temperature. Prints are laid emulsion upwards on the endless cloth belt and are pressed against the drum as it rotates. They emerge within 5 minutes, dry and well glazed. Prints not requiring a glaze may be dried emulsion downwards, then peeled carefully off the cloth as they emerge.

Rotational printing

Photography of virus bodies and other particles with facets of an equal dimension often results in some region of the body being less sharp or dense than another. This is overcome by reinforcing the image of one particular body during printing. The printing paper is held on a marked board and rotated about the axis of radially symmetrical along the axis of axially symmetrical objects. Exposures are then made of equal duration on each facet, until the paper has been rotated through 360° or moved along the whole axis of a structure. The developed image then shows a body of uniform density. This process is known as Rotational Printing Intensification (Markham). The dangers inherent in such intensification have been reviewed by Agraway, Kent & Mackay.

Density Masking

The photographic negative made in an electron microscop often varies in density due to uneven illumination and section thickness. A useful method for compensating variations in density makes use of a weak and highly defocussed positive held in register with the true negative. Prints are made from the combined negative. Generally plates with a high contrast are used to make the negative and correspondingly soft plates of the same manufacture are used to make the positive density mask (this is because they have the required contrast characteristics), e.g. if Ilford Special Lantern Contrasty plates are used for the negative, us Ilford Special Lantern Soft plates for the positive mask.

Defocus a suitable enlarger and then set at precisely 1:1 reproduction. Use a thin translucent sheet on top of the positive emulsion to ensure loss of all detail in the mask Place the soft plate exactly in the position of the negative image using a red filter over the enlarger lens and expose very briefly-a quick flash at f4.5 is sufficient. Develop th exposed plate with continuous agitation until, by visual inspection, the correct density is obtained. Process and dry the plate in the usual manner. Place the negative and its positive mask together so that both emulsions face outwards. This prevents images of the silver grains of the positive appearing on the print as they are out of the focal plane by two thicknesses of glass. Prints made by this method are virtually free of macro-contrast (differences in density between large areas). Micro-contrast or acutance (differences in density of details in the negative) is normally unchanged. Very occasionally it is improved to a small extent. This method is more reliable than hand shading in that areas normally inaccessible can be compensated for without trial and error. Light haloes may be formed around dense bodies, but these are usually dirt or stain crystals and haloes are rarely a disadvantage. Autoradiographs cannot be compensated for this reason.

Negative reversal

The replication method used in examining plant surfaces results in an initial photographic image that looks like a positive (although the plate or film may loosely be called a 'negative'). Before printing, the first negative must be contact printed against another plate or suitable cut film. This second image is a true negative of the original surface and when printed shows true dimensions and relief features of the surface and black shadows, making interpretation easy. When shadowed preparations are published or exhibited, they should be printed so that the black shadows fall from above. This aids interpretation. Where shadows are shown as white, i.e. without a reversal negative, they should rise from below.

Plates used for the contact printing may be similar to those used for the actual microscopy. Cut film, e.g. Ilford N8. 40 is often used, as this may be stored in the same envelope as the original negative and is easy to handle. Processing should be carried out as before.

Mosaics

It is often useful to build up a large picture of a specimen to compare tissue structures in a number of cells or to make calculations on the incidence of a particular feature in a number of adjacent cells.

Use the largest plate the instrument will take. Photograph the section at × 1,000 – 2,000, each negative overlapping the other by approximately one quarter to eliminate the barrel distortion of the objective lens at low magnification. Draw a sketch map of the relevant section and create co-ordinates on the map so that extra plates can be taken and matched with the original.

10

MEASUREMENT OF GROWTH

Growth is defined as increase in mass of living substance and is measured usually as increase in cell number, size or both. Often it is sufficient to measure only one of these parameters.

Ideally procedures for growth measurements of cell or organ cultures should have the following characteristics: (a) be applicable at any time without destroying the culture, (b) provide a true measure of increase in mass, and (c) be simple enough to be applied routinely.

A number of methods for measurement of growth have been described., Several, such as cell or nuclear counts, measurement of packed cell volume, or dry weight determinations, are direct methods. Others, such as measurement of nucleic acids, are indirect and depend upon known relationships to cell number or cell mass.

ENUMERATION OF CELLS AND NUCLEI

When cells can be grown as monodisperse suspensions or can be dispersed adequately and maintained as a stable suspension, cell counting techniques are preferred. For cells which are difficult to disperse the nuclear count technique is more precise.

A. Total Cell count: Haemocytometer

The haemocytometer count is the only feasible means of routine cell count in most laboratories. This method, however, is both time consuming and subject to substantial error. A single count, done carefully, may require 10-15 minutes and even under the best conditions, the method is subject to not less than 10% variation.

1. Clean a haemocytometer and coverglass by first rinsing in water, then in absolute ethanol and finally in acetone. Dry and polish with lens paper. Dry the coverglass with a piece of lint-free cloth.

2. Seat the coverglass firmly on the haemocytometer so that it covers both counting chambers.

3. Dilute 0. 5 ml of cell suspension to give a final concentration of approximately 1 x 10^5 - 2 x 10^5 cells/ml. If fewer than 100 or more than 200 cells are present in the chamber, the counting error is increased. Use of citric acid-crystal violet solution as a diluent minimizes clumping, stains the nuclei, and thus facilitates counting.

4. With a capillary pipette carefully fill both chambers of the haemocytometer. Be careful not to allow the fluid to overflow. Count the cells in each corner square and the center square of each chamber. The total number of cells counted in the ten squares x 1000 x the dilution previously made equals the number of cells per ml of the original suspension.

B. Total Cell Count: Electronic Counter

In an attempt to speed up red blood cell counting procedures and to increase precision, an electronic blood cell counter has been developed and made available commercially-:. This instrument is admirably suited to the counting of many animal cells grown in culture (l). The instrument operates on the principle of "electronic gating". A small aperature (100 44 in diameter and 75 4 long) is interposed between two large surface platinum electrodes which are immersed in electrolyte solution (saline). An electric current flows between the electrodes and through the orifice. By applying a vacuum, cells which are 'suspended in the saline on one side of the orifice are drawn through the orifice. As a cell passes through the orifice, being essentially a non-conductor, it causes a drop in the current which is picked up as an impulse, amplified and recorded on a decade counter. A mercury manometer with appropriately placed electric leads permits metering a sample of exactly 0. 5 ml. To eliminate backgraound count the threshold selector should be set so as just to include all cells in the population. A single count, including 1000 or more cells is completed in approximately 20 seconds. A 0. 5 ml sample of culture permits 5 or more successive counts on the same sample. The counting error is 1-2%.

1. Measure 24. 5 ml of 0. 85% NaCl (saline) into a 50 ml beaker. (This may be done quickly and accurately with a "Palo Pipettor"). Withdraw 0. 5 ml of sample to be counted and mix with the saline. This will yield a 1:50 dilution of the original sample.

2. Mix by pouring from one beaker to another 7-8 times.

3. Place the sample beaker on the platform of the electronic counter with the aperture tube and the electrode immersed in the suspension. The beaker should be positioned so that one side is against the aperture tube. This permits visual observation of the aperture with the microscope mounted on the counter.

4. Open the stopcock to the vacuum source allowing the mercury column to be displaced and the sample to flow through the aperture. Check with the microscope to be certain the aperture is clear.

5. With the appropriate switch, clear the decade counter and then close the stopcock to the vacuum source. As the mercury returns to normal level it will continue pulling sample through the aperture. Appropriately placed "leads" will activate and turn off the counter permitting the counting of exactly 0. 5 ml of sample.

6. Since the original dilution was 1:50 and the amount actually counted was 0. 5 ml the count recorded by the decade counter x 10^2 will equal the number of cells per ml of the original sample.

C. Viable Cell Count: Vital Stains

A number of "vital" staining procedures have been developed in an effort to differentiate between viable and non-viable cells. With some dyes, e.g., methylene blue or various tetrazolium salts, the dye is used to indicate metabolic activity although the relationship of metabolic activity to viability is not at all clear. With others, e.g., eosin Y, trypan blue and erythrosin B the ability of viable cells to exclude dye is the criterion used. Serious questions have been raised as to the validity of results obtained with vital stains. At best the results of such methods lack precision and must be interpreted with reservation.

1. Place 0. 5 ml of cell suspension (diluted to contain 1×10^5 - 2×10^5 cells/ml) in a 12×75 mm tube. Add 0.1 ml of 0.4% erythrosin B , and mix thoroughly. Allow to stand for 5 minutes but not more than 15 minutes.

2. With a capillary pipette, fill a haemocytometer as for cell counting.

3. Make a total cell count and a count of unstained cells. Assuming those cells which are unstained to be viable, express the results as viable cells.

* 0.4 gms erythrosin B in 100 ml of BSS.

D. Viable Cell Count: Plating

Another method for determining the number of viable cells is based on the ability of individual cells to multiply and give rise to colonies which can be enumerated. It gives a relative measure of viable cells. Media and conditions of growth have not been worked out to make the method feasible for many cell cultures. It is also too tedious to be useful as a rapid, routine technique except in isolated instances.

E. Enumeration of Cell Nuclei

Cells which are resistant to dispersion can be enumerated by the method of nuclear counts. The technique involves treatment with citric acid-crystal violet to destroy the cytoplasm and stain the free nuclei. The cells are then counted using a haemocytometer. The additional steps of handling and time involved make the method rather cumbersome for routine counting. Moreover, if a significant number of multinucleate cells are present, an obvious error is introduced.

MEASUREMENT OF PACKED CELL VOLUME (PCV)

Growth of cells in suspension culture, or in some cases in monolayer culture, can be measured conveniently by determining the packed cell volume of a given sample of culture. Increases in PCV can be used, therefore, to calculate the mean cell volume. In measuring PCV it is very important to use a constant gravitational force from experiment to experiment.

A. Hopkins Tube

The Hopkins tube is a graduated tube 16 mm in diameter, with a total capacity of 10 ml and having a capillary stem with a capacity of 0.05 ml calibrated in 0.01 ml divisions. Five or ten ml of sample is placed in the tube which is then centrifuged at 600 G in a bucket type centrifuge for 15 minutes. An angle-head centrifuge will slant the pellet and make it difficult to read accurately. The packed cell mass can be read directly. The chief disadvantage of the Hopkins tube is that the graduations are relatively coarse and rather large samples are required to get accurate readings.

B. Van Allen Hematocrit or Thrombocytocrit Tubes

The use of the Van Allen hematocrit tube for measurement of PCV has been described by Waymouth. It has the advantage that the tubes are readily available and are easy to use. The exact volume of suspension to be centrifuged is not measured in ml but a graduated portion of the tube is filled and following centrifugation the PCV is read directly as a percentage of the sample. The chief disadvantage of the method is that a rather heavy suspension is required to get a cell pack which can be accurately measured. The thrombocytocrit tube offers some advantage in this regard as it not only has a larger capacity but. the stem is narrower and has finer graduations. To use the thrombocytocrit tube a measured volume is filled into the bulb of the tube as with the Hopkins tube and following centrifugation the volume is read directly from the calibrations on the stem. Both the hematocrit and the thrombocytocrit tubes require spring clip closures.

MEASUREMENT OF DRY WEIGHT

A useful device for following the in vitro development of some cell populations is change in total cell mass as determined by changes in dry weight. To be meaningful this must be done with cultures in which the cells readily may be freed of medium components so that the dry weight, as determined, is a true measure of the weight of the cells.

Increase in dry weight indicates a net increase in cell mass. Coupled with measurement of packed cell volume and cell number it makes possible a calculation of net synthesis and changes in degree of hydration at both the cell and population levels. The chief limitations to the method are that it is time consuming and that rather large samples are required for accurate determinations. A precaution to be taken is that all samples be dried to a constant weight under specified conditions.

1. Aseptically remove a sample of the cells to be weighed to a conical centrifuge tube. The size of the sample will depend upon the density of the suspension and can vary from 1 to 25 ml. Wash with BSS (not more than three times) and resuspend in 1 ml of BSS. Excessive washing will lead to leakage of materials from the cells.

2. Transfer the sample quantitatively to a previously weighed aluminum planchet and dry to constant weight over P_2O_5. The drying process may be speeded by, (a) prior lyophilization of the specimen, or (b) use of a drying oven in place of P_2O_5.

CHEMICAL METHODS

The close relationship of protein and nucleic acid synthesis to cell growth has led many investigators to utilize serial measurement of these components as indicators of growth in vitro . Variations in the mean content of protein, RNA and DNA are such that no one measurement can serve as a reliable measure of population development. Use of one or more of these determinations in conjunction with direct methods, as outlined above, provides a much more reliable index of growth. The principle chemical methods employed are given below.

A. Extraction of Nucleic Acids and Proteins

Numerous methods for the extraction of nucleic acids and proteins have been proposed during the past few years. The subject of determination of nucleic acids in biological materials was reviewed by Hutchinson and Munro who concluded that the method of choice is a modified Schmidt Thannhauser procedure. Such a method is outlined. It should be emphasized that while this method is suitable for analytical purposes, it is not adequate for preparation of biologically active nucleic acids. For such purposes a phenol extraction method usually is employed.

B. Methods of Analysis

1. Protein

a. Micro-Kjeldahl Method

Since many proteins have similar nitrogen levels, the micro-Kjeldahl technique often is the method of choice. The technique consists of the conversion of nitrogen to ammonia and titration with a standard acid. Preparation of a standard curve permits conversion of the nitrogen values obtained to protein units.

b. Nessler Reaction

Although the micro-Kjeldahl method is extremely accurate, this analysis is much more time consuming than other methods available., The method to be described below is Nesslerization. As in the Kjeldahl procedure, the method utilizes the digestion of a sample and conversion to ammonium ions. However, the ammonia is subsequently complexed with a mercury reagent (Nessler' s solution) and the endpoint read spectrophotometrically. This method is fairly accurate through the range of 20 to 300 μg of nitrogen.

Materials

1. Sample of protein (containing about 100 μg nitrogen).
2. Standard ammonium sulfate (containing exactly 100 μg nitrogen).
3. Concentrated sulfuric acid (low in nitrogen), reagent grade.
4. Hydrogen peroxide, 30%, reagent grade.
5. Nessler' s reagent-.
6. Digestion tubes, pyrex (Folin-Wu tubes, 20 x 200 mm,

FLOW SHEET FOR THE SEPARATION OF RNA, DNA AND PROTEIN IN MAMMALIAN CELLS CULTURES IN VITRO.CELL PELLET (0.5 TO 2.0 × 10^6 CELLS)

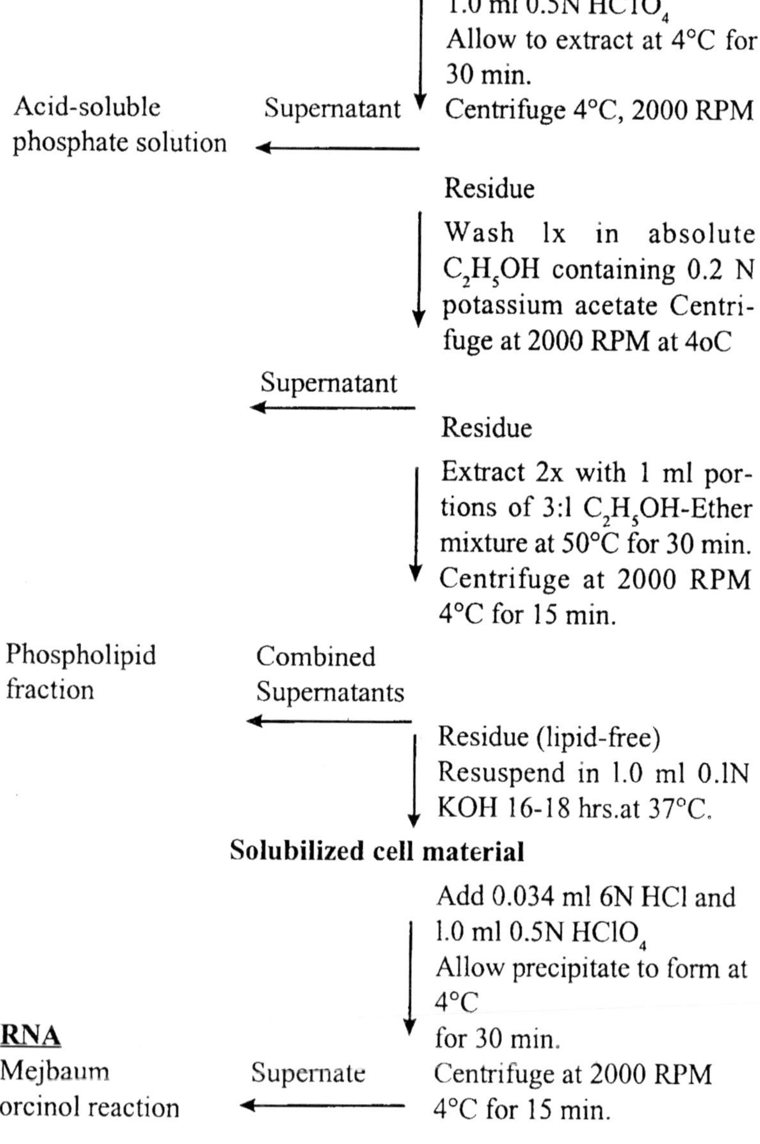

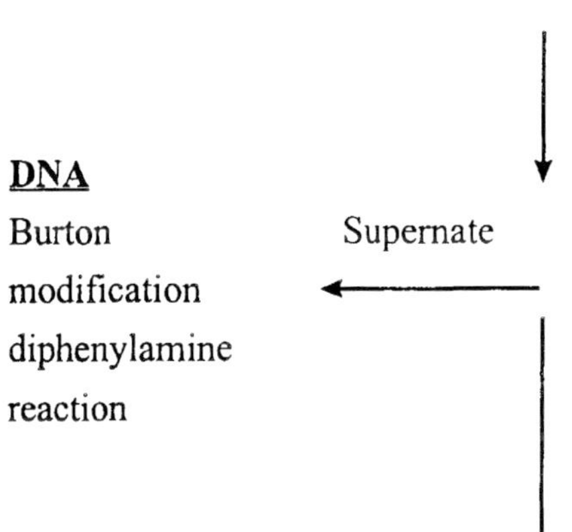

Residue - (DNA & Protein) Resuspend in 1.0 ml 0.5N $HClO_4$ Heat 90°C for 15 min.

Cool and centrifuge at 2000 RPM 4^0 C for 15 min.

Supernate →

DNA

Burton modification diphenylamine reaction

Residue- (Protein) Resuspend in 2.0 ml IN NaOH (may require overnight incubation at room temperature to put protein into solution).

Protein Solution →

Protein

Lowry modification Folin-Ciocalteau reaction

graduated at 35 ml and 50 ml).

7. Measuring pipettes, 1 ml.
8. Blunted stirring rods (7 mm x 25 cm)
9. Spectrophotometer
10. Metal digestion racks for Folin-Wu tubes

Procedures

1. Pipette duplicate 1 to 2 ml samples of unknown protein solution, reference standards and blanks (diluent for unknown sample) into Folin- Wu tubes.
2. Add 0. 5 ml of concentrated H_2SO_4 to each tube.
3. In a well-ventilated hood, heat tubes over a low burner for

15 min. to 20 min. When water has evaporated, turn flame up and continue to heat. The tubes will release dense fumes of sulfuric acid. Continue to heat for another 5 to 10 minutes.

4. Remove tubes from heat and cool for one minute.
5. Add one drop of H_2O_2 and reheat tubes over a low flame for 10 to 15 minutes. If color persists after this heating, allow tubes to cool, add another drop of H_2O_2 and reheat over a low flame for 5 more minutes. Finally, heat tubes over a higher flame for 10 to 15 minutes.
6. Allow tubes to cool.
7. Add distilled water to the 35 ml mark and place a blunted stirring rod in each tube.
8. With constant stirring, add 5 ml of Nessler's reagent to each tube. Permit reaction to work for 30 minutes.
9. Read optical density on spectrophotometer at 440 mμ in order in which Nessler's reagent was added.
10. Using a reference standard containing exactly 100 μg of nitrogen the mg of protein

$$= \frac{(\text{OD of unknown} - \text{OD of blank})}{(\text{OD of standard} - \text{OD of blank})} \times 0.1 \times 6.25.$$

the "protein factor" and assumes that animal protein contains 16% nitrogen.

C. Lowry Modification of the Folin-Ciocalteau Method

The determination of protein content by this method is based on the color reaction of the aromatic amino acids tyrosine and tryptophane with the Folin-Ciocalteau phenol reagent. The method assumes a constant tyrosine and tryptophane content in protein. The Lowry modification is presented here.

Materials

1. Sample to be tested (in lN NaOH)
2. Sodium carbonate, reagent grade
3. $CuSO_4 \cdot 5H_2O$, reagent grade

4. Potassium tartrate, reagent grade
5. Folin-Ciocalteau reagent
6. Phenolphthalein
7. Crystalline bovine serum albumin
8. Beckman DU spectrophotometer

Procedure

1. Add 0.5 ml of 110 $CuSO_4 \cdot 5H_2O$ to 0.5 ml of 2% potassium tartrate. Add this mixture, with careful stirring to, 50 ml of 2% Na_2CO_2. This is Reagent A.
2. Add 0.5 ml of an appropriately diluted sample to 5 ml of Reagent A. Incubate at 25°C for 10 minutes.
3. Titrate an aliquot of the Folin-Ciocalteau reagent to the phenolphthalein endpoint with 1N NaOH and on the basis of this titration dilute the reagent to 1N with distilled water.
4. Rapidly add 0. 5 ml of the IN Folin reagent to the 5. 5 ml of sample plus Reagent A. Mix uniformly as it is added.
5. At 30 minutes read the blue color which develops in a Beckman DU spectrophotometer at 750 mμ.
6. Construct a standard curve using crystalline bovine serum albumin over the range of 0-300 .g.
7. Read the values of protein of the unknown samples from this curve.
8. For greater accuracy the data can be analyzed by linear correlation method.

RIBONUCLEIC ACID

A widely used method for the quantitative determination of ribonucleic acid (RNA) involves measurement of the pentose content by the Orcinol reaction. This depends upon adequate separation from interfering substances. A modification of the Mejbaum reaction (19) is presented here.

Materials

1. Sample to be tested (fractionated by a method which insures separation from interfering substances while preserving the RNA).
2. Purified yeast RNA in phosphate buffer at pH 6.8.
3. Ferric ammonium sulfate, reagent grade.
4. Crystalline orcinol (recrystallized from Benzene)
5. Concentrated HCl (12N), reagent grade
6. Beckman DU spectrophotometer.

Procedure

1. Dissolve 13.5 grams of ferric ammonium sulfate and 20 grams of recrystallized orcinol in 500 ml of doubled distilled water. Store this as a stock solution in the cold (4°C).
2. To prepare a working orcinol reagent add 5 ml of orcinol solution to 85 ml of concentrated HCl and bring to 100 ml with doubled distilled H_2O.
3. Add 1 ml of sample (diluted to contain the equivalent of 4-40 4g of pentose) to 3 ml of the working orcinol reagent and heat in a boiling water bath for 20 minutes.
4. Cool and read the absorption of the green color which develops in a Beckman DU spectrophotometer at 670 mμ.
5. Construct a standard curve from data obtained using quantities of yeast RNA ranging from 0-100 μg/ml of aqueous solution.
6. Read the values of the RNA in the unknown samples from the standard curve.
7. For greater accuracy the data can be analysed by linear correlation methods.

DEOXYRIBONUCLEIC ACID

Quantitative determination of deoxyribonucleic acid (DNA) is most commonly done by measurement of deoxyribose by the diphenylamine reaction originally described by Dische and modified by Burton.

Materials

1. Sample to be tested (extracted by a method which will insure separation from interfering substances while preserving the DNA).
2. Diphenylamine (twice recrystallized from 70% ethanol.)
3. Redistilled glacial acetic acid, reagent grade.
4. Concentrated H_2SO_4, reagent grade.
5. Acetaldehyde, reagent grade
6. Crystalline herring sperm DNA
7. Beckman DU Spectrophotometer

Procedure

1. Diphenylamine stock solution is prepared by adding 1. 5 grams of diphenylamine and 1. 5 ml of concentrated H2S04 to 100 ml of glacial acetic acid. This stock may be stored at 4°C for several weeks.
2. Just before use add 0. 1 ml of acetaldehyde to 20 ml of the diphenylamine stock solution.
3. Add 1 ml of sample (containing 50-500 µg of DNA) to 2 ml of diphenylamine reagent and incubate at 37°C for 14-18 hours). The incubation time should be constant from experiment to experiment
4. Read the blue color which develops in a Beckman DU spectrophotometer at 600 mµ.
5. Using 0-100 µg of herring sperm DNA in aqueous solution construct a standard curve. Read the DNA values of the unknown samples from this curve.
6. For greater accuracy the data can be analysed by linear correlation methods.

MITOTIC COEFFICIENT

In many instances, particularly with explant cultures, it is not possible to make serial cell number measurements for following growth of cultures. In such cases it may be helpful to determine the mitotic coefficient. Practically, this is done by determining the number of mitoses per 100 cells. Such a method is dependent upon the ability to examine the culture with phase contrast optics at high magnification, or the culture alternatively must be fixed and stained. A comparison of the mitotic coefficients of cultures under varying experimental conditions often is useful. Where the mitotic rate is low, prior treatment with colchicine may be necessary. This procedure will yield the proportion of cells in a population undergoing division in a given unit of time and is known as the mitotic index (see glossary) .

RADIOAUTOGRAPHY

With increased availability of labeled compounds and with simplification of procedures the technique of radioauto-graphy has become a valuable and critical tool for the study of cellular metabolism and for the localization of inter- and intracellular organelles. Cell cultures provide an ideal starting material for investigating the synthesis and the localization of proteins and nucleic acids and for the effect of various agents (virus, irradiation, etc.) on cell metabolism.

The following provides a general description of the method. The choice of isotope, compounds, cells and media will depend upon the needs of the investigator.

Materials

1. Stock monolayer culture of cells on coverglasses.
2. Sterile screw-cap 16 × 125 mm tubes each containing one 9 × 22 mm #l coverglass.
3. Sterile cotton plugged 1 ml serological pipettes.
4. Sterile media containing isotope (e.g. 1μc/ml tritiated cytidine-uridine; 1μc/ml tritiated amino acid; 0. 3 4c/ml tritiated thymidine).
5. Sterile fine forceps.
6. Sterile balanced salt solution (BSS)
7. Columbia staining jars.
8. Glacial acetic acid: 95% ethanol (l:3) fixative.

9. Clean microscope slides (1 × 3 inches).
10. NTB3 Liquid Emulsion".
11. Water bath regulated at 42-45°C.
12. Dark room.
13. Wratten series #l Safelight filter with 15 watt bulb.
14. Coplin jars.
15. Light-tight black Bakelite microscope slide box.
16. D-11 developer.
17. Acid fixer with hardener'.
18. 0.25% Toluidine Blue stain (pH 6).
19. Canada balsam or Euparol.

Procedure

1. A monolayer coverglass culture in mid-logarithmic growth phase should be used.
2. Decant off the medium and add one ml of fresh medium containing isotope.
3. After various periods (minutes or hours) of incubation at 35°C, remove the coverglasses from tubes with fine forceps. Carefully rinse each coverglass in four changes of warm BSS and fix for forty minutes in the acetic acid: alcohol fixative. Remove the coverglasses from fixative and allow to air dry. For pulse labeling, the isotope-containing medium may be rinsed off with fresh medium and the culture allowed to continue for varying periods of time.
4. Mount the coverglasses for each time period on a clean microscope slide with the cells facing up. Use a minimum of Euparol as the mounting medium. Keep the coverglasses toward one end of the slide. [Where controls are required, such as ribonuclease, cut the coverglasses in half by lightly scratching the surface with a diamond pointed pencil and then break it into two equal halves. Incubate one half of each control coverglass in ribonuclease (1 mg/ml distilled water, pH 6. 5-7.0) for one

hour at room temperature. Rinse in water and allow to air dry.]

5. With a diamond pencil, label opposite side of slide with name and time of incubation and place it in a drying oven (45°C) for 24 hours.
6. Coat the slide with NTB3 liquid emulsion in the dark room. A sample of the stock emulsion is melted by incubating the bottle in a 42-45°C water bath for approximately 45 minutes. Check melted emulsion for bubbles and scoop out any that stand on surface. The dry slides should be arranged back to back in a coplin jar and then the jar placed in the water bath to pre-warm the slides. The dry slides, two at a time, are dipped once into the melted emulsion for 4- 5 seconds. Withdraw slides and allow to drain in a vertical position for a moment over the emulsion. Separate the slides, wipe the back of the slide with a piece of tissue paper and dry in a upright position. (This takes about one and a half hours. The safelight should be off during the drying period)
7. When slides are dry, transfer to a light-tight slide box, seal with black tape and leave box standing on edge (slides exposed horizontally) at room temperature for at least 48 hours. Periods up to six months may, be necessary.
8. Develop the slides in the dark room as follows:
   a. two minutes in D- 11 developer
   b. rinse in water
   c. fix in acid fixer for Z-5 minutes
   d. rinse in running water for 20 minutes (lights may be on now)
   e. rinse briefly in distilled water
   f. allow to dry and store for future staining
9. Stain slides in 0. 25% toluidine blue, rinse in 95% cthanol and air dry. Slides may be sealed with a coverglass using Canada balsam or Euparol as mounting medium.

The interpretation of the radioautography will depend on the isotope used, the time of incubation and the method of analysis. For example, if tritiated thymidine is used to investigate DNA synthesis, one may estimate the premitotic non- synthetic period (G2) by determining the time required for labeled DNA to show up in metaphase chromosomes. The grain count over metaphase cells will increase with time after the addition of label, being zero until the first cell passes through the G2 period and then will increase in number for a period of time after which there will be no further increase. The time required from the first signs of grains to where no further increase occurs will approximate the DNA synthetic period (S).

The length of time which cells spend in mitosis (M period) may be ascertained by the following formula

$$^{t}\text{mitosis} = \frac{MT}{0.693}$$

where M × 100 is ,the mitotic index and T is the generation time, assuming the culture is completely asynchronous (i.e. mitotic index is constant with time). The G_1 can be then determined knowing that generation time is the sum of G_1, S, G_2 and M periods.

INCREASE IN AREA

In following the growth of explant cultures or of some types of organ cultures, a useful measurement is the increase in surface area. This is usually determined as the increasc in surface area of a plane surface. It is of limited value since no consideration is given to increase in volume.

A. Camera Lucida and Planimeter

A camera lucida is a device which fits over the ocular of a monocular microscope and projects the image from the ocular onto a plane surface where the outline may be traced. The planimeter is a device which measures the area of any plane surface by moving a pointer around its boundary and reading the indicator on a scale.

B. Ocular Micrometer

A more rapid and convenient method, though less precise, is the use of a calibrated ocular micrometer. The diameter of a colony or explant may be measured in 2 or more directions in one plane. Taking the mean value of these measurements, one can calculate roughly surface area from the formula for the area of a circle.

MEASUREMENT OF CELL SIZE

Direct measurement of cell size can be accomplished by two techniques. A calibrated ocular micrometer can be used to measure mean cell diameter from which cell volume can be calculated. It involves the basic assumption that the cells, when free in suspension, are spherical. The "Coulter Electronic Cell Counter and Cell Size Analyzer" gives a direct measure of volume. An indirect measure of mean cell volume can be obtained by dividing packed cell volume by total cell number.

In any measure of cell size it is important to appreciate that suspending fluids, washing solutions, etc. , may cause changes in cell size because of osmotic phenomena. It is desirable, whenever possible, to suspend the cells in the growth medium for measurement. If other suspending solutions are used appropriate controls are necessary.

A. Ocular Micrometer

1. Transfer a sample of the cells to a 12 × 75 mm tube. Dilute the cells, in growth medium, to give a concentration of approximately 5×10^5 cells/ml.
2. Place a drop of the cell suspension on a clean microslide and cover with a clean coverglass. Do not compress
3. Bring the cells into focus with the 40 X objective and measure the diamcter of each of 100 cells using a calibrated ocular micrometer.
4. Applying the formula 3rrr3, calculate the mean volume of the cells.

B. Electronic Counter

1. Measure 49 ml of growth medium or saline into a 50 ml beaker. Add 1 ml of cell suspension. (Saline may be used providing significant changes in cell size do not occur when the cells are transferred from growth medium.)
2. Pour the sample from one 50 ml beaker to another 7-8 times to obtain a uniform distribution of cells in the sample.
3. Place the beaker on the platform of the Coulter Counter and make a total cell count at a threshold level which will include all of the cells in the sample.
4. Raise the threshold 5 or 10 divisions and make another count. Repeat at 5- 10 division intervals. Stir carefully between counts to avoid settling of cells.
5. Determine the differences in count between each two successive threshold settings. This represents the increment of the population falling in this particular size range. Plot these values against threshold settings. Data can be also plotted as percentage of the total count as a function of the increments between successive threshold settings.
6. To obtain absolute values for cell size the counter must be calibrated with particles of known volume, e.g. , erythrocyte, polystyrene particles'

C. Mean Cell Volume

Measure packed cell volume (PCV) as described already. Prior to centrifuging the sample make a total cell count. Divide the PCV by the total cells (cells/ml × volume) to obtain a measure of mean cell volume.

11

CELL FRACIONATION

You have already learnt about the existence of cell organelles. You may, however, be left with the impression that intracellular structure is an artifact of fixation and staining or of the optical system used. Isolation of cell organelles in a functional state from homogenized cell masses provides dramatic evidence of organelle reality. The isolated organelles may then be tested for enzyme content to demonstrate the intracellular compartmentalization of metabolic pathways. Rat liver will be used in this exercise because it provides a rather uniform population of cells which are easily homogenized. Enough material can be obtained from one rat for an entire labora

tory class. Sucrose solutions are used because physical and biochemical properties of the isolated organelles are retained. Equipment and solutions used in the fractionation technique should be kept cold (0 to 5°C) to slow protein denaturation and thus maintain organelle morphology and biochemical activity.

When the cell organelles are viewed microscopically outside the living cell their existence can no longer be doubted. The identity of these organelles, however, will still be a problem to the beginning student. The rat liver cell is typically about 30 m in diameter. Nuclei are typically 12 to 15 μm in diameter, and because of their DNA content, stain easily with methyl green. The rat liver cell contains approximately 2,500 mitochondria which are typically 1 to 3 m rods. Several other particles such as lipid droplets also occur in this size range in liver homogenates. The mitochondria can be identified because of their specific enzyme content. One convenient enzyme for assay is succinic dehydrogenase which oxidizes succinate to fumarate. Using a tetrazolium-succinate reagent, electrons from succinate are transferred to the tetrazolium dye thus reducing it to an insoluble formazan, as follows:

succinate +tetrazolium (soluble yellow)

succinic dehydrogenase ↓

fumarate + formazan (insoluble purple)

The use of isolated cell fractions has provided our best evidence for cellular compartmentalization of biochemical activities. Cell fractions will be used to demonstrate the compartmentalization of the electron transport pathway.

THEORY OF CENTRIFUGATION

Definition of symbols

F = force

m = mass

a = acceleration

ω = angular velocity

r = distance of particle from center of rotation

RCF = relative centrifugal force

g = acceleration due to gravity

rpm = revolutions per minute

T_S = time for a particle to sediment

s – sedimentation coefficient (Svedbergs)

d = diameter of particle

η =viscosity of medium

δ_p = density of particle

δ_m = density of medium

In subsequent formulae subscript 1 indicated start of centrifugation, and subscript 2 indicated end of centrifugation.

Centrifugal Force Equation

The centrifuge is an instrument used to apply force to particles in suspension. In general,

$$F = ma$$

but in a centrifuge the acceleration is applied radially.

$$F = m\omega^2 r$$

Converting mass to weight, w

$$F = \frac{w}{g}\omega^2 r$$

It is most convenient to refer to the relative centrifugal force when comparing forces between different centrifuges. If the weight of the particle equals 1, RCF gives multiples of the particle's weight in the earth's gravitational field.

$$\text{RCF} = \frac{\omega^2 r}{g}$$

but,

$$\omega = \text{rpm} \times \frac{2\pi}{60}$$

$$\text{RCF} = \frac{\text{rpm} \times 2\pi}{60} \times \frac{r}{980}$$

$$= 1{,}118 \times 10^{-8} \times r \times \text{rpm}^2$$

Sedimentation Time Equation

The time necessary to centrifuge a suspended particle to the bottom of a test tube can be calculated from the following formula:

$$T_s = \frac{1}{s}\frac{\text{In}\, r_2 - \text{In}\, r_1}{\omega^2}$$

but

$$s = \frac{d^2(\delta_p - \delta_m)}{18\eta}$$

$$T_s = \frac{18\eta(\text{In}\, r_2 - \text{In}\, r_1)}{\omega^2 d^2(\delta_p - \delta_m)}$$

Sedimentation Velocity Centrifugation

Centrifuge tubes are filled with a cell homogenate and spun in a fixed angle rotor until the largest particles have pelleted at the bottom of the tubes. During centrifugation, the particles are driven diagonally across the tube and slide down the centrifugal wall. In these fixed angle rotors, sedimentation distance is shorter than it would be if the particles were to sediment through the entire length of the centrifuge tube. The fixed angle rotor thus has the advantage of reducing the sedimentation time.

From Figure 11.1 we can see that the pellet contains all of the largest (fastest sedimenting) particles, as well as a significant number of smaller particles. The smaller particles in the pellet were those near the centrifugal wall at the beginning of centrifugation and did not have far to sediment in the tube. Inability to purify large particles is a distinct disadvantage of the sedimentation velocity method. Only the smallest particles, which will be the last ones in suspension, can be isolated in pure form

In practice, sedimentation. velocity centrifugation is a very useful technique. Much has been learned about biochemical content of cellular organelles by measuring composition or activity of pellets relative to whole homogenate.

Density Gradient Centrifugation

Centrifuge tubes are partially filled with layers of solution which varies in concentration and therefore density. The most dense layer is placed at the bottom of the tube and successive layers of decreasing concentration (density) added on top of this. In *sedimentation density gradient centrifugation,* a small volume of cell homogenate is layered on top of the gradient. During centrifugation, organelles will sediment downward into the gradient until they come to a layer whose density is equal to their own. When all the organelles have reached their isodensity layer, the system is at equilibrium and should contain discrete layers of organelles at different density levels in the centrifuge tube. Layers can be removed from the gradient with a hypodermic syringe

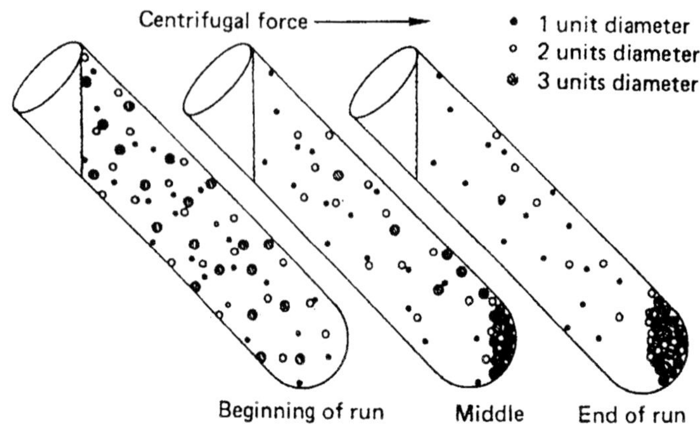

Figure 11.1 : *Sedimentation velocity centrifugation*

or by punching a small hole in the bottom of the centrifuge tube and collecting drop fractions.

Density gradient centrifugation should provide essentially pure organelle fractions. Optimum separation of particles is not always achieved with *sedimentation density gradients* because of aggregation. Aggregation produces a zone of high-density cell material on top of a low- density solution. This unstable situation results in visible droplets of aggregated material streaming into the gradient. Droplets often do not disaggregate during the centrifuge run. As a result, organelle layers are impure. Ways to overcome this artifact include using fewer grams of cells per volume of homogenate or using a *flotation density gradient procedure.* This procedure involves making the homogenate dense enough to layer in the gradient below the isodensity layer of most organelles. In the course of centrifugation, aggregates will remain low in the gradient and only free unaggregated organelles will float up to their isodensity layer.

Density gradient centrifugation has the disadvantage of being limited to small volumes of material. The advantage of the technique is that extremely pure organelle fractions can usually be isolated. Cell organelles have a very narrow density range (about 1.05 to 1.35 g/ml) and often several organelles will overlap in density.

Peroxisomes, lysosomes, and mitochondria are often close to 1.15 g/ml in density. If density gradient centrifugation fails to achieve fractionation, a third procedure called *zone centrifugation* may be used.

Zone Centrifugation

Sedimentation velocity centrifugation demonstrates that larger particles move faster in a centrifugal field. Density gradient centrifugation demonstrates that a layer of particles can be moved by centrifugal force into a clean layer of solution. By applying both of these principles, a mixture of particles can be fractionated, even if they differ only slightly in size.

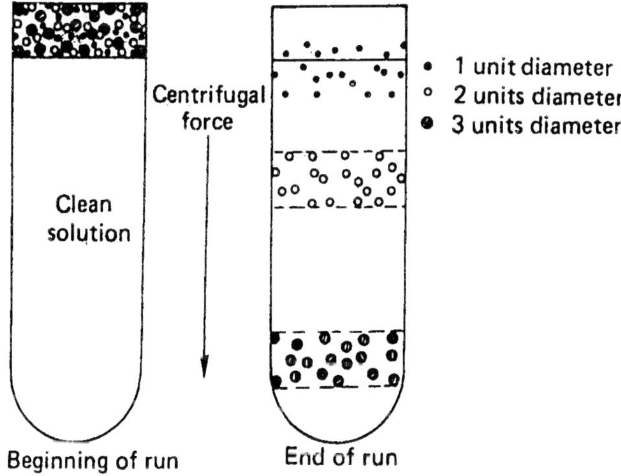

Figure 11.2 : ***Zone centrifugation***

A layer of homogenate can be placed on top of clean solution (this does not even have to be a gradient). This tube is centrifuged not to equilibrium, but only long enough for different size particles to separate. If particles differ in diameter by a factor of only two, they will sediment with a *velocity* difference of four. A band or zone of organelles with the same diameter will sediment at the same velocity through the medium. Zones of organelles with different diameters will consequently separate during sedimentation.

Zone centrifugation is rapid, and, with special rotors, can fractionate large amounts of material. This method is becoming a standard preparative technique in physiology laboratories.

METHOD

Preparation of 10% Rat Liver Homogenate

Before killing the rat, place the following items in an ice bath

1. 150-200-ml beaker containing about 50 ml of a 0.5 M sucrose solution
2. tissue press
3. tissue homogenizer
4. 100-ml graduated cylinder
5. eight 50-ml plastic centrifuge tubes
6. two 400-ml beakers
7. cold 0.5 M sucrose solution

Kill a healthy adult rat. While the liver is being removed, weigh the cold sucrose-containing beaker to the nearest 0.1 g. Place the liver in the beaker, reweigh, and calculate the weight of the liver to the nearest 0.1 g. Measure out enough cold 0.5 M sucrose in the cold 100-ml, graduated cylinder to prepare a 10% homogenate (9 ml sucrose to 1 g liver). Pour some of this sucrose into a cold 400-ml beaker. Transfer the liver to this beaker and mince with scissors. Alternatively, the liver can be macerated into the beaker with a tissue press. Pour an aliquot of this slurry into the cold tissue homogenizer. Work the pestle into the homogenizer until most of the sucrose solution is forced above the pestle. Draw the pestle forcefully back so that the sucrose is sucked into the tip of the homogenizer with maximum turbulence. This turbulence creates tremendous shear forces which break the cell membranes. Continue homogenizing in this fashion until the cell mass has completely liquified; then do about three more up-and-down strokes to assure maximum cell breakage. Pour this homogenate into the second cold beaker and continue homogenizing aliquots of the slurry. When the entire liver has been homogenized, add the remainder of the measured volume of sucrose solution and mix thoroughly.

Divide this 10% rat liver homogenate into three portions for use in the following parts of this exercise. Save 15 ml for use in *Density Gradient Centrifugation*. Save 1 ml for *Microscope Observations*, which are described in Part IV. The remainder of the 10% homogenate is used for *Sedimentation Velocity Centrifugation*.

Density Gradient Centrifugation

Immediately start preparing a high-density homogenate by adding 8 g sucrose to 8 ml of 10% homogenate. Allow this mixture to sit at room temperature with occasional stirring. Preparation of functional cellular fractions by density gradient centrifugation should, of course, be done at 0 to 5°C. However, at room temperature the tetrazolium-succinate reaction will occur during the course of centrifugation. Morphology and density of the organelles will remain fairly typical for this short period of time.

The following procedure is designed for the Spinco SW 25.1 swinging bucket rotor. However, fractionation of the homogenate can be obtained with a tabletop centrifuge equipped with a swinging bucket rotor. (Thirty minutes at 3000 X g will produce visible separation of organelles.) Number three cellulose tubes for the Spinco SW 25.1 rotor. Place 1 ml of tetrazolium-succinate in the bottom of tube 3. Using a hypodermic syringe without a needle, place 4 ml of room-temperature 2.6 *M* sucrose in the bottom of each centrifuge tube. The remainder of the three gradients is to be prepared by adding each of the following solutions in order of decreasing concentration: 1.8 *M*, 1.6 *M*, 1.3 *M*, and 0.7 M sucrose. Draw 4 ml of the room-temperature sucrose solution into a syringe. Attach an 18- or 20-gauge hypodermic needle. Trickle the sucrose solution down the side of each tilted centrifuge tube so that the solution hits the meniscus at sucrose), layered carefully at the bottom of each gradient. This can be done using a syringe with a long hypodermic needle which can be lowered into the sucrose gradient. Ease in a drop of high-density homogenate and let this drop settle to its isodensity layer. Raise or lower the tip of the needle to this layer and slowly inject the remainder of the homogenate. Mark the boundary between sucrose layers on all three tubes with a felt-tip pen or a wax marking pencil so that density boundaries can be identified at the completion of the run. At this time, be sure to observe the large aggregates or droplets settling from the homogenate into the clear sucrose gradient in tube 1. Droplet formation should be seen in this sedimentation density gradient, but *not* in the two flotation density gradients. Load the Spinco SW 25.1 swinging bucket rotor with the three gradients and centrifuge at 20,000 rpm for 30 minutes at room temperature. Place the three gradient tubes in a test-tube rack. Make observations and drawings indicated in the *Analysis* section. When observations have been completed, density bands can be sampled with a hypodermic syringe, or by collecting drop fractions. an oblique angle and flows over the surface of the denser layer already in the tube. When the three gradients are prepared, layer 3 ml of homogenate on top of gradient 1. The second and third gradients are to have 3 ml of the high-density homogenate (8 ml homogenate plus 8 g).

MICROSCOPE OBSERVATIONS

Place a small drop of the 10% rat liver homogenate, resuspended nuclear and mitochondrial pellets, supernatant II and any fractions taken from density gradients on separate microscope slides. Place a cover slip on each drop and press it flat under a towel or cloth. This wet mount slide will provide a minimum depth of field and minimum light diffraction so that cellular structures can be most easily identified. It will be useful to line up this series of slides on separate microscopes in the laboratory. You can then move easily from one microscope to another to compare purity and do particle counts of the different sedimentation velocity and density gradient fractions.

Nuclei can easily be identified by their staining with methyl green. One drop of rat liver fraction and one drop of methyl green on a wet mount slide should show staining in ten minutes. Mitochondria will have to be identified by examining the layer

TABLE 11.1

	Number per microscope field							
	Whole cells	**Nuclei**	**Red blood cells**	**Mitochondria**	**Lipid droplets**	**Lysosomes**	**Perioxi somes**	**Other**
10% rat liver homogenate								
Washed nuclear pellet								
Mitochondrial pellet								
Supernatant II								

TABLE 11.2

Number per microscope field								
				Number per microscope field				
	Colour	**Density range**	**Membrane fragments (present or not)**	**Nuclei**	**Red blood cells**	**Mitochondria**	**Lipid dropl-ets**	**Lysosomes peroxis-omes**
Tube 1								
Band 1								
2								
3								
4								
5								
6								
7								
Tube 2								
Band 1								
2								
3								
4								
5								
6								
7								

Sedimentation Velocity Centrifugation

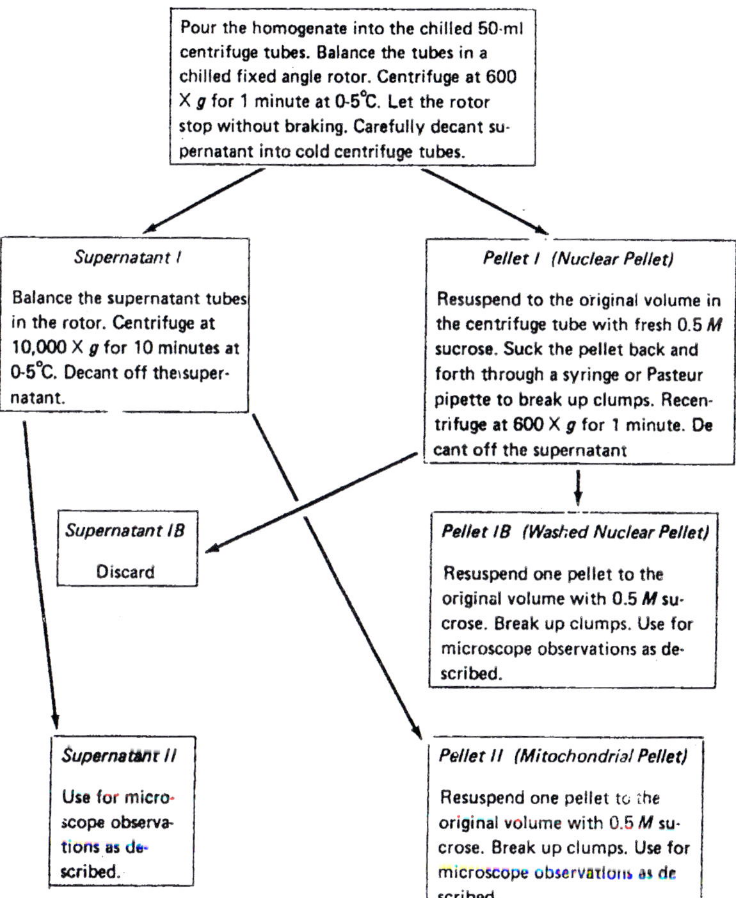

in density-gradient tube 2 which corresponds to the purple layer in tube 3. Protein denaturation accompanying staining prevents microscope identification of mitochondria in tube 3.

Starting with the original 10% homogenate, you should measure diameters of and be able to identify intact rat liver cells, nuclei, nucleoli, mitochondria, red blood cells and lipid droplets. Where will lipid droplets be found? Remember, their density will be less than 1.00. If a very good microscope is available, microbodies (peroxisomes), lysosomes and broken endoplasmic reticulum may be seen. Do the mitochondria and nuclei of intact and ruptured cells differ in size? Is the number of nucleoli per nucleus constant?

ANALYSIS

Review of Centrifugation Theory

State how sedimentation time will be affected by:

1. increasing the viscosity of the medium.
2. shortening the centrifuge tube.
3. changing from a swinging bucket to a fixed angle rotor.
4. increasing the diameter of a particle by a factor of 2; by a factor of 10.
5. increasing the density of the medium until the density is greater than the density of the particle.

Sedimentation Velocity Centrifugation

Drawings of the liver cells and organelles may be useful for future reference. Be sure to indicate the size in micrometers. It is important to get some idea of purity of fractions prepared by successive centrifugation steps. Determine the number of each kind of particle in a typical microscope field from the original homogenate, the two pellets and the final supernatant. Place the counts in Table 11.1 and compare.

12

THE ELECTRON MICROSCOPE

We derive our information about the world, in the first instance, mainly by looking at it. Sight is our dominant sense, and our sensory world is primarily a visual one. Magnifying glasses, microscopes and telescopes provide extensions of our visual sense, enabling us to see, or see more clearly, small or distant objects. The invention and gradual perfection of these instruments, permitting the exploration of regions of the world inaccessible to our unaided eyes, is a major and well-known theme in the history of science. Much of the progress in biology in the last hundred years or so is closely bound up with improvements in the microscope. The discovery and study of bacteria and protozoa, the identification of cells as the fundamental constituents of plants and animals, the recognition of chromosomes and their role in heredity, the gradual increase in our understanding of the structure and development of tissues and organs-all these, to mention only some of the more obvious examples, were made possible only by the development of ever more perfect microscopes and the associated techniques of specimen preparation.

The electron microscope is one of the most recent stages in this development. It renders open to direct inspection a level of structure finer than any accessible before, and, just as did the light microscope, it reveals a new world, many of the features of which were previously unsuspected. As a result of the new observations which it has made possible, our understanding of the organization of plant and animal tissues has been enormously extended, and many of our ideas about the way cells are constructed and the way they function have been radically changed. The electron microscope has also added greatly to our knowledge of the structure and reproduction of viruses.

This booklet sets out to explain briefly the principles on which the electron microscope works, and to describe some of the biological discoveries which have been made with it.

RESOLVING POWER

In order to grasp the reasons which led to the development of the electron microscope it is first necessary to understand some of the limitations of the light microscope. This necessitates a rudimentary knowledge of microscope optics, and in particular of the meaning of the term *resolving power.*

The type of microscope with which the reader is most likely to be familiar is the ordinary compound microscope which uses visible light. This is referred to as an 'ordinary' microscope because there are several other kinds of microscope which also use visible light but are optically more complicated; the phase-contrast, interference and polarizing microscopes are the main examples. The ordinary light microscope-which will simply be called the light microscope from now on-consists essentially of a light source and three sets of lenses. The *condenser* focuses light on to the object, a magnified image of which is formed by the combined action of the *objective* and *eyepiece.*

In constructing the path of light rays through an optical system, use is made of what are called ray optics. That is, it is assumed that light travels in straight lines and undergoes refraction on entering a medium of different refractive index. Using ray optics it is possible, knowing the focal length of the various lenses and their mutual arrangement, to calculate the magnification of a microscope. This can be expressed, for present purposes, as a linear magnification, which is the ratio of the length of the final image to that of the object. The magnification of light microscopes usually lies between about × 25 at the lower end of the range, to × 1,500 at the upper.

The magnification of a microscope is easy to calculate but it is not, in itself, a particularly useful piece of information. The statement that the magnification of a microscope, using a certain combination of lenses, is say, × 400, tells us by how much the apparent size of the object will have been increased, but gives no information about a much more important characteristic, namely the amount of detail we may expect to be able to see. Our aim in using a microscope is, of course, to see more detail than we can make out with our unaided eyes. Magnification is only a means to this end, and it is of no value if the image finally produced contains no more detail than we could see-or *resolve-by* eye.

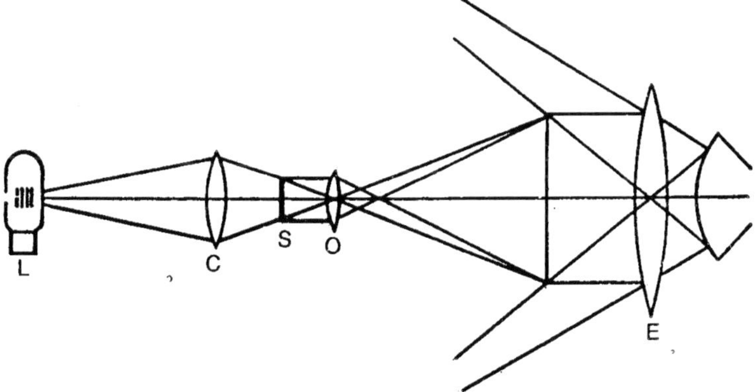

Figure 12.1 : ***Diagram showing in simplified form the path of rays throyugh the light microscope. Light the lamp (L) is focused on to the specimen (S) by the condenser (C). The image formed by the objective (O) is further magnified by the eyepiece (E).***

The fineness of detail which can be made out with a microscope is referred to as its *resolving power.* Suppose we are examining two small objects. Provided they are well separated from each other a microscope will form distinct images of them, but if they are gradually brought closer together a stage will eventually be reached at which the two images merge and the microscope will fail to distinguish them as separate. It is in these terms that resolving power can be defined: it is the smallest separation at which we can see that two objects are present rather than one. The smaller the resolving power, the greater the amount of detail that can be made out.

The factors which limit the resolving power of a microscope

cannot be understood in terms of ray optics. It is necessary instead to take account of the fact that light has wave properties, and to consider the way in which the image in a microscope is formed. What happens to light when it encounters an object in the microscope? This is a rather complicated topic, the theory of which was worked out in the second half of the nineteenth century, largely by Ernst Abbe. It is unnecessary to go into this theory in any detail here, but one or two basic points must be brought out, since they are relevant to an understanding of the electron microscope.

An object becomes visible in a microscope as a result of an interaction between it and the light waves used to illuminate it. This interaction, which causes a disturbance or deviation of the waves as they pass the object, is called diffraction. Light waves which do not interact with the object will not be diffracted. According to Abbe's theory, the detail in the final image arises as a result of interference between the diffracted and undiffracted light, which in some parts of the image plane will arrive in phase with each other and tend to reinforce, while in others they will arrive out of phase and tend to cancel each other out. This results in a pattern of light and dark areas, which is the image of the object. For a detailed exposition of this theory the reader must refer to books on the, theory of the microscope. For present purposes all we need do is consider in slightly more detail the question of the interaction between light waves and object. The essential point here is that light waves will be disturbed only by an object which is sufficiently large in relation to their wavelength. Very small objects will not bring about any detectable deviation in the waves, and will therefore remain invisible (or unresolved). It is not difficult to see that the smaller the wavelength of the light the smaller the object can be which will cause diffraction, and hence the better the resolving power will be. The resolving power of a microscope is, in fact, directly related to the wavelength of light used to illuminate the object.

This is an extremely important result. It means that the nature of light itself sets a limit to the amount of detail that can be resolved in a microscope. The reason why the upper limit in magnification of a microscope is set at about x 1,500 is that at that level we can comfortably see all the detail that we can hope to make out. Further magnification merely results in a larger, but no more informative image, and is analogous to examining a photograph in a newspaper with a magnifying glass in the hope of making out more detail. In the microscope it is the wavelength of light which limits the amount of detail; in the printed picture it is the number of dots to the square inch.

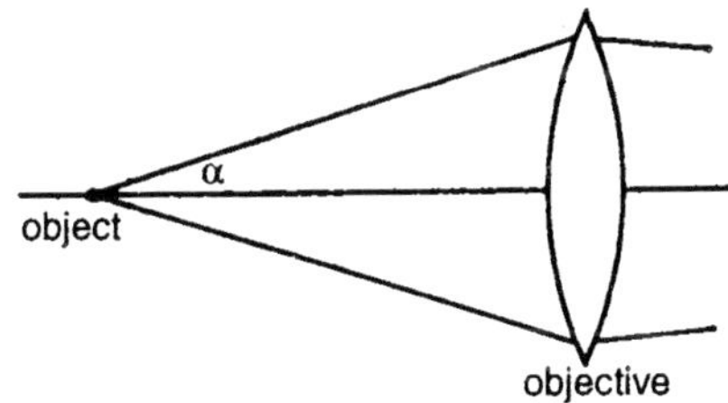

Figure 12.2 : *Diagram defining the semi-angle* (α) *subtended by the object at the objective lens.*

The explanation of image formation in terms of diffraction of light waves leads to another important result, which cannot be explained without giving a detailed account of the whole theory, and can only be stated here in general terms. It is that the resolving power of a microscope will depend not only on the wavelength of the illumination but also on the extent to which diffracted light is accepted by the objective lens. This in turn is related to the angle subtended at the front lens of the objective by the object. If this angle is small (as, for example, when the objective is far away from the specimen) then only a narrow cone of light is accepted by the lens and much of the diffracted light is lost. The resolving power is consequently poor. Conversely, when the angle is large, more of the diffracted light is admitted and the resolving power is improved. It is for this reason that high-power objective lenses have to be brought up very close to the specimen, while low-power ones have a large working distance.

Taking into account the two factors, wavelength (λ) and the semi-angle α subtended by the object at the objective, the resolving power (δ) can be expressed as a first approximation by the equation:

$$\delta = \frac{\lambda}{\sin \alpha}$$

To be exact, however, two further factors need to be taken into account. Firstly, the resolving power depends not only on the angle α but also on the refractive index (n) of the medium between the specimen and the objective. In most cases the medium is air, which leaves the equation unchanged, but the highest power objectives are used with oil or water filling the space between objective and specimen, and the resolving power is then reduced by I/n In the case of oil the value n is about 1.4. Secondly, for theoretical reasons which need not be entered into here, a constant, 0.5, has to be introduced into the equation, which now becomes:

$$\delta = \frac{0.5\lambda}{n \sin \alpha}$$

The quantity n sin α is termed the numerical aperture (N.A.) of the objective.

We can now insert some real values in the equation and calculate the resolving power. The unit of measurement most commonly used in light microscopy is the *micron (μ)*, which is 10^{-4} cm. Visible light has a wavelength of about 0.5 μ and the numerical aperture of the best lenses is about 1.4. From these two values the resolving power comes out as approximately 0.2 μ. This is the best resolving power that can be achieved with the ordinary microscope; the numerical aperture cannot be increased, and even using visible light of the shortest possible wavelength (about 0.45 μ) the resolving power is not significantly improved.

It should be clear that the only way to improve resolving power is to decrease the wavelength of the illumination. To make the effect of wavelength quite clear it may be helpful to consider a simple analogy. Suppose a blindfolded person were trying to explore a room simply by feeling around with a probe. The amount of detail he could detect would depend on the thickness of the instrument he used. With a walking stick he could make out only the grosser objects-the chairs and tables, the position of the door, and so on. To resolve finer detail-to discover whether the chairs were carved or plain, for example-it would be necessary to use a finer instrument, such as a pencil. With a needle it might be possible to make out the texture of the upholstery. The thickness of the probe, which here limits resolution, is the analogue of wavelength in the microscope.

Ultraviolet light, with a wavelength of about 0.3 μ, offers the possibility of an approximately twofold improvement in resolving power. Unfortunately, the practical difficulties of working with ultraviolet light are fairly considerable, since it is not significantly transmitted by glass. In consequence the lenses of the microscope, as well as the slides and coverglasses, have all to be made of quartz or fluorite, which are expensive materials and difficult to work. A special lamp is, of course, necessary, and since our eyes cannot detect ultraviolet radiation (except in so far as they are damaged by it) the image has to be recorded photographically. Under the best conditions the ultraviolet microscope has a resolving power of about 0.1 μ, and until the advent of the electron microscope this represented the limit of resolution. Subsequently, however, it has been entirely superseded as a high-resolution instrument. Its usefulness now lies in the fact that certain cell components (such as nucleic acids) selectively absorb ultraviolet light of characteristic wavelengths, and can thereby be identified and located in the cell. The study of the localization of different chemicals in cells is called cytochemistry, and the ultraviolet microscope is still an important tool in this kind of work.

For a really significant improvement in resolving power, however, it is necessary to turn from visible or ultraviolet light to some other form of radiation, and it is here that the development of the electron microscope begins.

Wave properties of electrons

The reader was probably introduced to electrons as particulate constituents of atoms, circling the nucleus in a series of concentric orbits. While electrons can certainly be treated as particles for some purposes, from the present point of view it is more important to know that in appropriate conditions they also display wave properties. In other words, they have some of the characteristics of other forms of wave motion (such as visible light, ultraviolet light, and X-rays) and can be treated theoretically in much the same way. In particular, electrons just like visible light, have a wavelength associated with them. The wave properties of electrons were predicted on theoretical grounds by the French physicist, de Broglie, in 1924, and were confirmed experimentally a few years later.

In the next section we shall describe how beams of electrons can be produced and what factors affect their wavelength. That discussion can be anticipated, however, by saying that the wavelength of electrons is always very much less than that of visible light, and that this opens the way to a correspondingly large improvement in resolving power.

Electron guns

The source of illumination in an electron microscope is called the electron gun. This usually consists of a small V-shaped piece of wire, the filament or cathode, together with two circular metal plates with holes drilled in their centres. A large voltage is applied between the filament (negative) and one of the plates, the anode (positive). A current flows through the filament and heats it to incandescence, causing it to emit electrons. These are attracted towards the anode, which is oppositely charged, and some of them pass through the hole in its centre. The proportion which does so is increased by the presence of the second metal plate, which is negatively charged with respect to the filament. It is called the cathode shield. Its negative charge has the effect of concentrating the electrons emitted by the filament into a beam passing symmetrically along the axis of the gun, and hence through the hole in the middle of the anode.

The property of emitting electrons when heated is common to all metals and is called *thermionic emission*. It is used in cathode-ray tubes, in valves, and in various other electronic devices. The higher the temperature of the filament the more electrons are emitted, and in practice it is usual to make the filament of tungsten, which can be heated to over 3,000° C without melting.

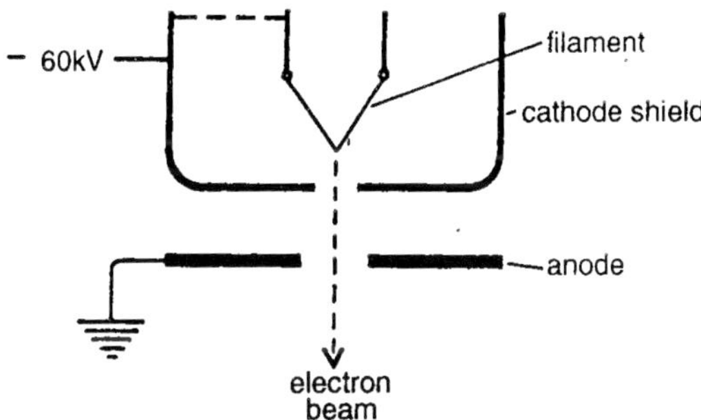

Figure 12.3 : *Diagram of an electron gun.*

The effect of the gun is to produce a narrow beam of electrons passing at high speed in a given direction. The voltage applied between the filament and the anode and which accelerates the electrons is usually between 40,000 and 100,000 volts (40—100 kV), though a few microscopes have been built which work at higher voltages than this. The greater the voltage, the greater the energy of the electrons. The accelerating voltage (V) also determines the wavelength of the electrons. The unit of length in which this is measured and which is commonly used in electron microscopy is not the micron (μ) but the Angstrom unit (Å), which equals 10^{-8} cm (I μ = 10^4 Å) The wavelength of electrons, in Å, is given by the approximate formula:

$$\lambda = \sqrt{\frac{150}{V}},$$

where V is in volts. Thus, using an accelerating voltage of 60 kV, the wave length will be $\sqrt{\frac{150}{60,000}}$, which equals 0.05 A. Comparison of this figure with the wavelength of visible light (0.5 μ or 5,000 Ã) shows that the effective wavelength of an electron beam is about 10,000 times less than that of a light wave. Other things being equal, this would lead to a corresponding increase in resolving power. However, as we shall see, there are many factors which prevent this theoretical degree of improvement being realized in practice and the actual gain in resolving power is very much smaller. Nevertheless, although the theoretical improvement cannot be achieved, the gain in resolution made possible by the short wavelength of electrons is still very great. The actual resolving power which has been obtained will be discussed after the other components of the electron microscope have been considered.

Electron lenses

The fact than an electron beam is deflected by a magnetic field has been known since the turn of the century, and has long been made use of in the cathode-ray tube. It was not until the 1920's, however, that it was demonstrated that a radially symmetrical magnetic field, such as is formed by a coil of wire with a current passing through it, will act as a lens and can be used to focus an electron beam.

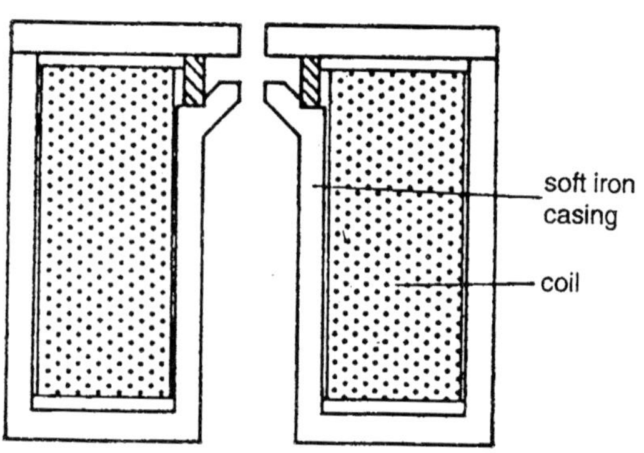

Figure 12.4 : *Diagram of an electron lens in section.*

The actual path of electrons in such a lens need not be considered in detail here. It can be deduced by application of the 'left-hand' rule, more commonly applied to electric motors, and in general it is found that the path of electrons is helical. By appropriate design a lens can be produced which will form a point image of a point source of electrons. In practice an electron lens consists basically of a lens coil, made up of a few thousand turns of wire, with a current of about I amp or less flowing through it. The magnetic field produced is concentrated by a soft iron casing around the coil and, in some cases, by other, specially shaped pieces of soft iron, called polepieces, in the centre of the coil. The focal length of such a lens depends on the value of the current flowing through it, and this can usually be varied. At maximum excitation of the coil the focal length of the lenses used in an electron microscope is usually a few millimetres.

SIMPLE ELECTRON MICROSCOPES

The electron gun and the electron lens just described together form the essential components of an electron microscope. In its general layout this is basically similar to a light microscope, except that it is inverted. The electron gun takes the place of the lamp and acts as the source of illumination, and the glass lenses of the light microscope are replaced by electron lenses of similar functions. Our eyes are not, of course, sensitive to electrons and so the final image is projected either on to a viewing screen coated with a material which fluoresces when irradiated with electrons, or on to a photographic plate if the image is to be recorded permanently. Since it projects the final image, the lens in an electron microscope which corresponds to the eyepiece of a light microscope is called the *projector lens*. The other lenses, *condenser* and *objective,* have the same names as in the light microscope and serve the same functions.

One extremely important property of an electron beam has not so far been mentioned. It is that it can be produced only in a vacuum. Electrons cannot travel more than a very short distance in air, since they are stopped by collision with gas molecules. The interior of an electron microscope must therefore be evacuated. The actual vacuum required is about 10^{-4} mm Hg, and this is obtained by means of vacuum pumps. As we shall see, the requirement that the microscope must work under vacuum imposes severe limitations on the kind of specimen that can be examined.

Simple microscopes of the kind were constructed experimentally

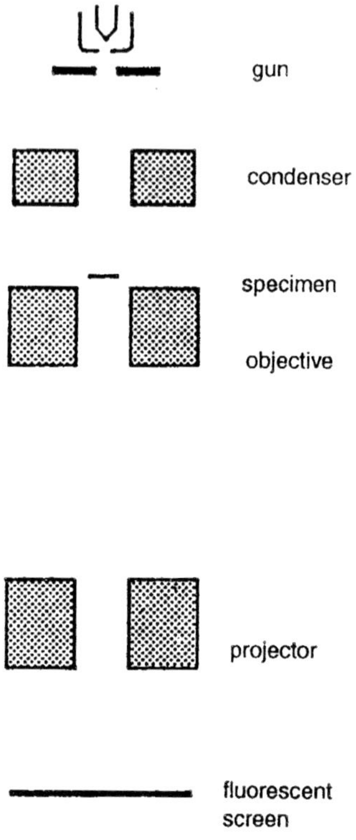

Figure 12.5 : *Diagram showing the layout of a simple electron microscope.*

in the 1930's, and it was soon shown that their resolving power could be better than that of a light microscope. These early instruments were primitive in design and difficult to operate. Nevertheless, they contained all the essential features of the modern electron microscope, and the development of the latter has been mainly a question of incorporating large numbers of technical refinements, which have led to a progressive improvement in performance and ease of operation. The basic principles have remained unaltered.

Modern electron microscopes

The layout of a typical high-resolution electron microscope. It differs from the early instruments chiefly in that there are more lenses: instead of one condenser lens there are two, and instead of there being just two lenses to form the image of the specimen there are four, additional *intermediate lenses* being inserted between the objective and projector. One reason for having two condenser lenses is that they make it possible to produce a narrower illuminating beam, and thereby limit the area of the specimen which is irradiated by electrons. This may be important in order to prevent damage to the specimen. The intermediate lenses allow the magnification to be changed over a wide range. In practice the range in magnification of an electron microscope is usually between about × 1,000 at the bottom end of the scale (overlapping with the light microscope) to about × 200,000 at the top. The very highest magnifications are not commonly used, however, since the maximum resolving power attainable with biological materials can usually be achieved with smaller enlargements.

The fluorescent screen on which the final image appears is viewed through thick glass windows in the chamber at the base of the microscope column. For critical focusing of the microscope, the image on this or another viewing screen can be examined through a low-power binocular microscope mounted outside the column. The screen can be tilted to permit this, and it can also be raised out of the way to allow photographs to be taken with the plate camera situated below. The whole microscope column, from the gun at the top to the plate camera below is about six feet high. It is usually fairly massively constructed, in order to ensure mechanical stability and prevent disturbance by vibration of the building, etc. The lens coils, through which moderately large currents are flowing, would tend to get hot and are therefore usually cooled by water circulating around them.

The various lenses and the gun all have to be accurately lined up on a common axis, and for this there are mechanical controls for moving them about. The specimen itself is moved on a special stage equipped with delicate controls operated by fine micrometer screws. With these the specimen can be moved accurately over small distances and placed in any desired position.

As already explained, the interior of the microscope has to be evacuated, and this requires high-performance vacuum pumps, as

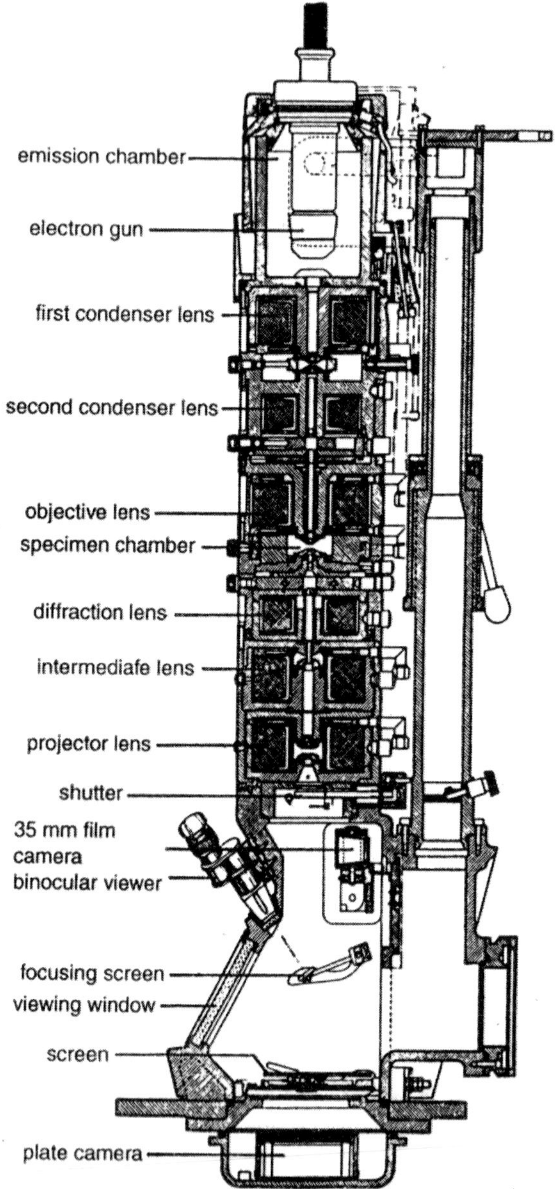

Figure 12.6 : ***Section of the column of a high-resolution electron microsocpe.***

well as a gauge to measure the pressure. The pumps are usually located in the desk on which the microscope column is mounted. In order to be able to change the specimen without breaking the vacuum in the column an air-lock is provided, through which the specimen can be introduced into the microscope with only a very small amount of air.

The other main components of the electron microscope are the electrical supplies, the most important of which are the high-voltage system for the electron gun and the circuits providing the lens currents. There are important reasons why these have to be fairly elaborate. All electron lenses suffer from defects. One of these is analogous to the fault found in many glass lenses and is called *chromatic aberration:* the focal length of the lens varies with the wavelength of the illumination. As already noted, the wavelength of an electron beam depends on the voltage used to accelerate the electrons, and if this voltage is not kept constant the wavelength will obviously vary. In practice the high-voltage supply has to be stabilized to about I in 100,000 in order to provide a sufficiently 'monochromatic' beam. The focal length of an electron lens also depends on the value of the current flowing in the coil, and this too must obviously be highly stabilized if the focal length is to remain constant. Again, the permissible variation is about I part in 100,000. Each lens is provided with a separate stabilizer. The precise value of each lens current can usually be controlled by means of resistances, and the lenses can also be switched off completely if necessary. The high-voltage supply of the gun and the filament-heater current can also be adjusted by electrical controls, so that an electron microscope usually has a fairly formidable control panel of knobs and switches. All the operations which in a light microscope are done mechanically, such as changing the objective or eyepiece to alter the magnification, or racking the body of the microscope up and down to bring the specimen into focus, are carried out in the electron microscope by adjusting electrical controls.

With its large column, numerous lenses, specimen stage, camera, pumping system and complicated electronic supplies the electron microscope has become a highly elaborate instrument, hardly recognizable as the offspring of the simple experimental microscopes made just over thirty years ago. Nevertheless, that is what it is. Beneath all the refinements and elaboration the basic principles remain unaltered, and from the point of view of understanding the role of the electron microscope in biological research it is these that matter.

It remains only to consider the resolving power of the modern electron microscope. As already noted, the wavelength of electrons is only a fraction of an Ångström unit, but the resolving power which this should theoretically make possible has not been achieved in practice. There are several reasons for this. One important one is that all electron lenses are highly imperfect compared with the glass lenses of the light microscope. Apart from their chromatic aberration, which has already been mentioned, they also suffer from spherical aberration, which means that their focusing action varies in passing from the optical axis outwards to the periphery. In the light microscope this defect can be largely overcome by combining lenses of different shape, which correct each other's faults, but this approach is not yet possible with electron lenses. The solution which has to be adopted (and which is also used to some extent in the light microscope), is to work with only the central portion of the lens, and this is done by building into it an aperture of small diameter. The restriction in the aperture of the lens cannot be carried too far, however, without deterioration of image quality arising from other sources, and in practice a compromise value has to be adopted. Spherical aberration in the lenses is one of the main obstacles preventing the attainment of the theoretical resolving power, but it is not the only one, and a number of other factors are involved. For example, as the resolving power is improved it becomes necessary to make a corresponding improvement in the mechanical stability of the microscope. Obviously, if we are hoping to achieve a resolving power of a few Ångström units the specimen must not move by that amount while it is under observation. Ensuring this can be a matter of considerable difficulty, since we are dealing here with distances comparable to those which separate atoms in molecules or crystals. The column and specimen stage have to be specially designed to eliminate vibration and movement, and precautions have to be taken to keep the specimen at a stable temperature, otherwise thermal expansion will cause it to move about.

The resolving power which is actually achieved depends largely on the quality of design and construction of the microscope, and also on the skill and experience of the operator. It also depends on the nature of the specimen being examined. The best results which have been obtained so far come from work on certain crystals, in which it has been possible to resolve spacings in the crystal lattices of just over 2 Å. With biological specimens the resolving power is not at present as good as this, and is more likely to be in the region of 10 Å. However, compared with the resolving power of the light microscope—0.2 μ or 2,000 Å—this represents an improvement of about 200 times, which is adequate for most purposes at present.

OTHER TYPES OF ELECTRON MICROSCOPE

The type of microscope which has just been described is a conventional, high-resolution instrument of the kind which is most commonly used in biological laboratories. A number of smaller instruments, of lower resolving power (20-100 Å) are also available, and are used for work not requiring the best possible performance. These have fewer lenses and less elaborate electronics. In addition to these, however, there are various special types of microscope which make use of electron beams and electron lenses but are otherwise considerably different in their method of operation. A detailed description of these lies outside the scope of this booklet, but one example which may be mentioned is the scanning microscope, which is chiefly used for examining the surfaces of solid objects. In this instrument a fine beam of electrons is made to scan the specimen, just as an electron beam scans the screen in a television tube, and an image is built up sequentially. Depending on the chemical composition and topography of the surface of the specimen, electrons are reflected or absorbed to different extents in different regions. The reflected electrons can be collected and made to form an image on a fluorescent screen by suitable electronic devices. The resolving power of such microscopes is limited by the size of the electron beam used to scan the specimen, and is rarely better than 100 Å. For examination of surfaces, however, this is an extremely useful type of microscope.

13

VIRUSES AND MOLECULES

The development of the light microscope as a tool of biological research which went on in the eighteenth and nineteenth centuries, depended not only on improvements in the microscope itself but also on the invention of increasingly sophisticated techniques for preparing material for examination. At the outset the early microscopists were largely restricted to examining small objects that scarcely required any preparation. They studied protozoa or other minute organisms from pond water, looked at insect mouthparts, examined pollen, feathers, hair and butterfly scales. All these could be placed under the microscope intact. To use the microscope to study the internal structure of larger organisms required greater ingenuity. One of the first methods to be developed, and which was used a great deal at one time, involved soaking small pieces of tissue in solutions which dissolved the material binding the cells together, causing them to fall apart. This process was called maceration. With favourable material it could yield useful information about the shapes of cells, and even some details of their internal organization. Another simple technique consisted of making smears or squashes of small pieces of tissue. Provided the cells separate or flatten sufficiently, a great deal of their internal structure can be made out in such preparations. Squashes are still commonly used for studying chromosomes. These methods, however, will work with only some kinds of material, and are of rather limited scope when it comes to elucidating in detail the three-dimensional structure of tissues and organs. This requires a technique which preserves the original shapes and mutual relationships of the cells, and yet permits these to be seen. The only way to do this is to cut the material into a series of thin sections, and methods for doing this were worked out in the nineteenth century. Concurrently with the development of all these techniques for producing specimens small enough and thin enough to be studied there was a great deal of research into methods for preserving cells and tissues, so that they would withstand subsequent processing, and for staining them, so as to make their structure more clearly visible.

Details of some of the methods used in light microscopy will be touched upon below, where they are relevant. What is important to note now is that the light microscope became a really useful tool for the biologist only when the problems of preparing material for examination had been satisfactorily worked out; for precisely the same thing happened when the electron microscope became available. The preparation of material in suitable form presented even greater problems than it did with the light microscope, and a great deal of effort and ingenuity had to be expended to devise satisfactory techniques. For reasons which will become clear later, it is important that at least the outlines of these techniques should be understood, and they will be discussed in some detail in the following pages, in which the applications of the electron microscope in biological research are described. The first objects which will be considered are viruses, which were among the earliest biological specimens to be examined in the electron microscope.

PREPARATIVE TECHNIQUES FOR VIRUSES

All viruses are exceedingly small, but some are smaller than others. A few of the larger ones can just be seen in the light microscope as minute dots, but their shape cannot be made out and nothing can be seen of their internal structure. Before the advent of the electron microscope the shapes and dimensions of viruses had to be deduced indirectly; for example, by determining how fine a filter they would pass through, or by studying their rate of sedimentation in a centrifuge. Such methods could give only very approximate results. Almost nothing was known about the internal structure of viruses. Before considering how the electron microscope has changed this situation we must describe the methods by which viruses are prepared for examination.

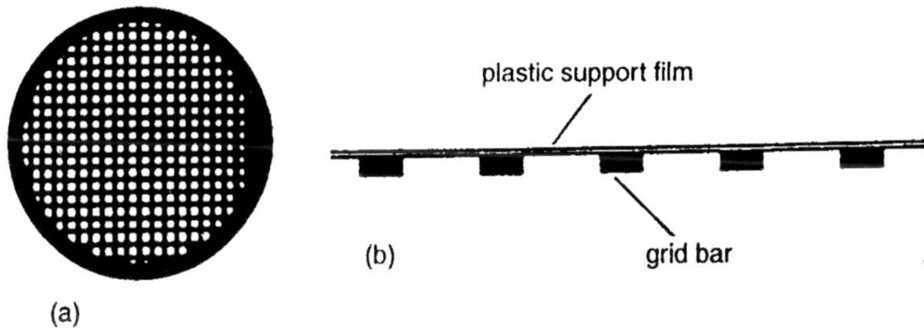

Figure 13.1 : ***(a) A grid for supporting specimens (actual diameter 3 mm) (b) part of a grid and plastic support film, in section.***

Specimens which are to be studied in the electron microscope have to be supported on something, and this obviously cannot be a glass slide such as is used in the light microscope, since this would be opaque to electrons. Instead of a slide use is made of a fine metal grid or mesh, 2-3 mm in diameter and usually made of copper, with holes about 0.1 mm across. (Grids of this kind are usually made by electrolytic deposition on a suitably shaped cathode.) The grid is covered before use with a thin plastic film, commonly of nitrocellulose (celloidin), which forms the actual support for the specimen. Such films are prepared by dissolving the plastic in a volatile solvent and allowing a drop of the solution to spread on a water surface; the solvent evaporates, leaving behind a sheet of plastic only a few hundred Ångström units thick. This is thin enough to be transparent to electrons, yet strong enough to withstand irradiation.

Viruses are usually obtained from infected plants or animals by grinding up pieces of tissue. This has to be done vigorously, in order to break up the cells in which the viruses multiply, and it is usually done in a special piece of apparatus called a homogenizer. The technique is considered more fully in the next chapter. Alternatively, it is sometimes possible, to obtain viruses from the body fluids of infected organisms. However obtained, the suspension or fluid can then either be examined without further treatment, in which case the virus particles will occur mixed up with all sorts of cell debris and other material, or else steps can be taken to extract the viruses and free them from contamination. In outline, the latter process commonly involves the use of a centrifuge to spin out the heavier cell fragments, etc., leaving the virus particles in suspension. The latter can then be concentrated by centrifuging at higher speed, and they can be washed if necessary by resuspending them in a suitable saline solution and centrifuging again. Many technical variations are possible in this procedure, but the end result aimed at is always a pure preparation of virus particles, uncontaminated by foreign material. Small drops of the suspension of virus particles may then be placed on plastic-covered specimen grids, and allowed to dry. Often the suspension is sprayed on to the grids through an atomizer, which produces conveniently small droplets.

If such a preparation were examined in the electron microscope without further treatment the results would almost certainly be disappointing. The virus particles would probably be seen, but they would be indistinct, and scarcely stand out against the background. It would almost certainly not be possible to make out any internal structure in them. The reason for this is important, since it applies not only to viruses but to biological specimens in general. We therefore have to discuss now why the image of such virus particles should be so poor in contrast and lacking in detail, and to describe what must be done in order to improve it. To do this we must leave viruses for the moment and consider how an image is formed in the electron microscope.

To understand image formation it is simplest to revert to considering electrons as particles. An electron beam consists of a stream of such particles, accelerated down the microscope column by the high voltage of the electron gun. As already explained, electrons are readily stopped or diverted by collision with atoms or molecules, which is the reason for working in a vacuum and also for supporting the specimen on an extremely thin plastic film. The specimen itself is, of course, made of atoms and will therefore also act as an obstacle to the passage of electrons. In passing through the specimen some of the electrons in the beam will encounter atoms, and as a result of the collision will be diverted or scattered. If they are scattered through a sufficiently large angle they will be lost to the electron beam, since all the lenses of the microscope are fitted with narrow apertures. Other electrons, however, will pass straight through the specimen and will end up on the viewing screen or photographic plate. The extent to which electrons will be scattered by any particular part of the specimen depends on two things: the number of atoms per unit volume, and the size of the atoms in question. The more atoms there are in a given region, and the larger they are, the greater is the probability that an electron passing though will collide with one of them. The dark areas in the final image therefore correspond to regions in the specimen which scatter large numbers of electrons, thereby removing them from the beam, while the light areas correspond to less dense regions which allow electrons to pass freely. The distribution of densities in the final image is closely related to the distribution of matter in the specimen.

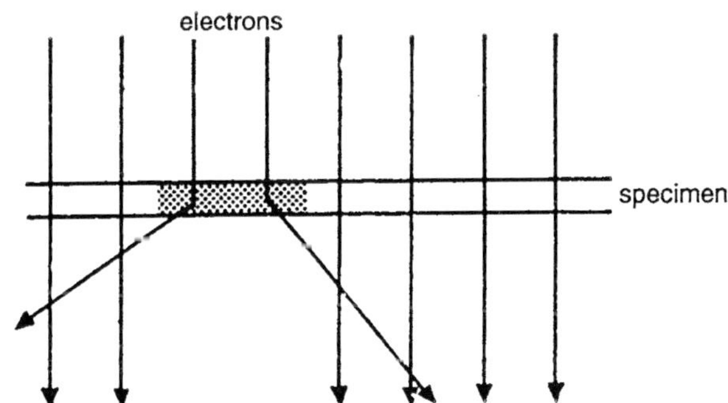

Figure 13.2 : ***Diagram to illustrate how electrons are scattered in passing through a dense region of the specimen (stippled), but not by less dense regions.***

The size of atoms can conveniently be expressed in terms of the atomic number of the element in question. Elements with high atomic numbers are highly effective in scattering electrons, those with low ones are not. This fact explains why it would have been possible to make out very little detail in a preparation of viruses made simply by allowing the particles to dry down on a support film; viruses, like living organisms in general, are made up almost entirely of light elements, such as carbon, hydrogen, oxygen, nitrogen, phosphorus and sulphur.

In order to obtain satisfactory images of viruses or other biological specimens in the electron microscope it is necessary to treat them in some way so as to enhance their electron-scattering power. This is analogous to staining cells or tissues for examination in the light microscope. In 'staining' for the electron microscope we make use not of dyes, which selectively absorb light of certain wavelengths, but of certain salts of heavy metals, such as lead, uranium or tungsten, which have high atomic numbers. In the simplest kind of procedure the material to be examined is soaked in a solution of the stain (for example, uranyl acetate) and, if it combines with this sufficiently, its visibility in the electron microscope will be increased. The success of this kind of staining method depends on both the specimen and the stain. Proteins, however, which are present in all living organisms, quite readily combine with salts of heavy metals and it is therefore usually not difficult to achieve some degree of staining.

The simple and direct method of staining just described is used a good deal for cells and tissues, as will be explained in the next chapter, but it is not much used for viruses, since there are other, more effective techniques for making their shape and structure visible in the electron microscope. These also make use of the electron-scattering power of heavy metals, but do so in different ways. In the simplest method the stain is used not to impregnate the virus particles but to form a thick, opaque layer around them. The virus particles, being largely unstained, then stand out as light objects against a dark background. This is called 'negative staining'. The method is simple in practice, since it is only necessary to mix the virus particles with the solution of stain and to spray the mixture on to grids covered with support films. Uranyl acetate is one of the substances commonly used as the 'stain' in this method, and so are sodium or potassium phosphotungstate. These substances dry out to form almost structureless solids; they surround the virus particles very closely and fill all the crevices and hollows in their surfaces, so that a great deal of minute structural detail may be revealed, as we shall, see.

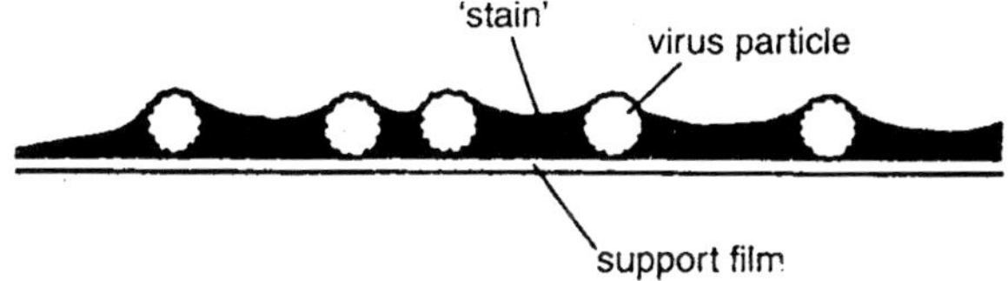

Figure 13.3 : ***Diagram to illustrate the technique of negative staining, in which the specimen is made to stantd out against an electron-opagque (dark) background.***

The second technique commonly used for studying viruses was developed earlier than the negative-staining method, but is somewhat more complicated. In this case heavy metals are again used as electron-scattering materials, but in the form of the metal rather than as salts. The technique is called 'shadowing'. A preparation of 'virus particles is made on a grid in the usual way, without staining, and then a thin layer of metal is deposited over it. This

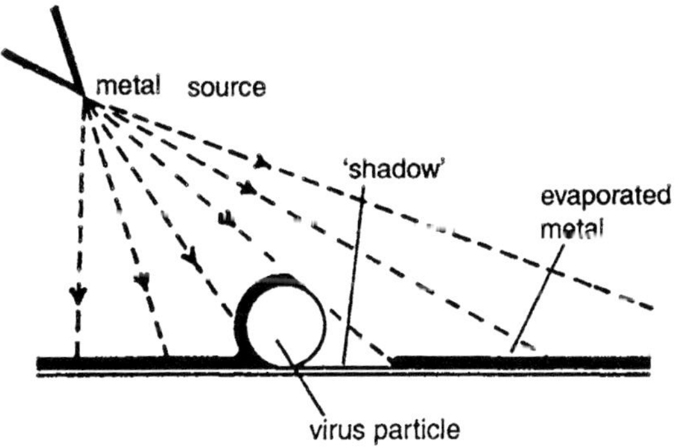

Figure 13.4 : ***Diagram illustrating the shadowing technique.***

is done by evaporation: a piece of metal wire is heated to incandescence, in a vacuum chamber, with the specimen placed nearby. If the specimen is placed not directly facing the metal source but at an angle to it, the metal will not be deposited in the 'shadow' of objects (such as virus particles) standing up above the level of the support film. The shadowing technique does not reveal internal structure in viruses, but it can give fairly precise information about their shape and dimensions. This is particularly the case if care is taken to determine the angle of shadowing accurately, since the length of the shadow then enables the height of the specimen to be calculated.

The two techniques-negative staining and shadowing-illustrate the important point that most biological specimens are of too low electronscattering power to be studied in their original form in the electron microscope, and that the enhancement of contrast depends on the use of heavy metals. Further aspects of specimen preparation will be considered in the next chapter, when the structure of cells and tissues is considered. The remainder of this chapter will be devoted mainly to describing some of the facts which the electron microscope has revealed about virus structure.

VIRUSES

The electron microscope quickly yielded a large amount of new information when applied to viruses. At the outset many different sorts of viruses were examined and it was found that each had a characteristic shape and size. Many were approximately spherical, ranging in diameter from not much more than 100 Å to almost 0.5 μ in a few cases. Others were rod-shaped, and some (the viruses attacking bacteria) were more complicated in form. The examination and description of virus particles of different kinds was a necessary preliminary to more detailed experimental work on their chemical composition, and on the processes by which they are produced inside living cells. Research on viruses, just as on any other kind of organism, depends on their accurate description and characterization in the first place, since without this it may be impossible for the observations of one worker to be repeated by another. Taxonomy is still a fundamental part of biology, even at the level of viruses.

Apart from enabling the shapes and sizes of different kinds of viruses to be described, the electron microscope also provided an accurate method for counting virus particles. In quantitative experiments on virus growth it is usually necessary to be able to measure the concentration of virus particles, and in many cases this is not easy unless direct counting is possible. In one of the simpler methods for counting viruses the suspension containing them is mixed with a known concentration of polystyrene latex particles, which are small plastic spheres about the same size as virus particles, and the mixture is then sprayed on to grids with an atomizer. The contents of each droplet dry down separately and the ratio of virus to latex particles is estimated from counts made in the electron microscope. Since the concentration of latex particles is known, that of the virus can be calculated.

The electron microscope, then, has enabled us to examine and count virus particles. It has done more than this, however, for it has also been an indispensable tool in recent research which has extended our understanding of viruses down to the molecular level. The structure of some viruses can now be described to a considerable extent in terms of the different types of molecules of which they are made and the way these molecules are fitted together, and a good deal is also known about the manner in which new virus particles are formed inside cells. It is essential to note, however, that although the electron microscope has played an important part in this work, it has been only one among many methods of investigation. The detailed picture of virus organization which is now emerging owes just as much to biochemical and genetical studies, and to immunological work, as it does to observations with the electron microscope, and much of the information provided by the electron microscope would be difficult to interpret without the results obtained using other techniques. In the remainder of this section we shall confine ourselves mainly to a discussion of some of the principal features of virus structure, rather than their growth in cells, since it is here that the contribution of the electron microscope has been particularly important.

It is known from chemical analyses of purified preparations that most virus particles consist entirely of protein and nucleic acid. Except in a few special cases no other type of molecule is present. The first question we have to consider, therefore, concerns the way in which the protein and nucleic acid are arranged in the virus particles. Here it is possible to make an important generalization.

In spite of their diversity in shape and size, all viruses prove to be constructed on an essentially similar plan : the protein forms an outer case or shell, and the nucleic acid is inside. This is shown by various pieces of evidence. With some viruses it is possible to separate the protein from the nucleic acid simply by rapidly diluting the medium, which subjects the particles to 'osmotic shock'. After doing this it is found that the virus particles look empty in the electron microscope, instead of having dense contents. Chemical analysis shows that their nucleic acid has been liberated into the medium, while the empty particles consist of protein alone. It has also been found that the nucleic acid of most viruses is protected from attack by enzymes as long as the particles are intact, but becomes susceptible if the particles are damaged. There is evidence from various sources that, of the two components, it is the nucleic acid which is essential for the reproduction of the virus, since it is infective by itself, whereas the protein is not. The protein can therefore be regarded as forming a protective shell around the nucleic acid, preventing it from damage by the environment.

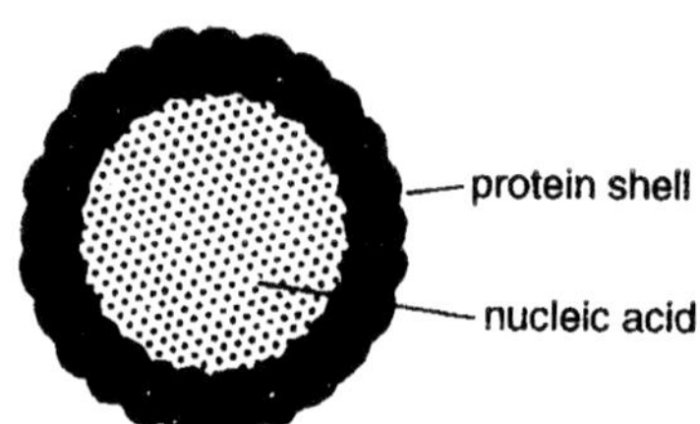

Figure 13.5 : *Diagram illustrating schematically the basic structure of a virus.*

Figure 13.5 shows an electronmicrograph of one of the spherical viruses, human wart virus, prepared by the negative-staining technique. As already noted, this technique shows up surface structures. It will be seen that the particles do not have smooth surfaces but are apparently made up of small globules, or subunits. These are outlined by the stain, which has penetrated into the grooves between them. We know that in this micrograph we are seeing the outer surface of the virus particles, and are therefore looking at the protein coat. What, then, do the globules represent? The answer is that they must be either single protein molecules or small aggregates of them. Several lines of evidence point to this conclusion. First, it agrees with what is known about the dimensions of protein molecules. Although proteins are long-chain molecules, made up of large numbers of amino acids, they do not usually occur in completely extended form. The polypeptide chain is commonly folded up in an intricate manner, giving rise to a molecule which is approximately globular in overall shape. Such globular molecules are usually a few tens of Ångström units in diameter. Second, in the case of one virus (tobacco mosaic virus, Plate Ia and b) it has been possible to make a very detailed study of the protein coat, using a variety of methods in addition to electron microscopy, and it has been shown that the molecules which form it are ellipsoidal in shape, about 60 Å long and 25 Å in diameter, and that these arc regularly and closely packed in a layer one molecule thick. Such detailed information about the size and shape and arrangement of the protein is not available for most other viruses, but the appearance of more or less globular subunits in negatively stained preparations of many kinds of virus particles suggests that the basic arrangement is probably very similar.

For the time being, since it is not certain in most cases whether the globules we see in electronmicrographs of virus particles are single protein molecules (as in tobacco mosaic virus) or small aggregates of them, the non-committal term 'subunit' is used. It will bc noted in Plate IC that the subunits appear to be regularly arranged; that is, they are not scattered at random but are evenly spaced over the surface of the particlcs. This is found to be the case with all viruses of which sufficiently good electronmicrographs have been obtained. Careful analysis of the micrographs shows that the pattern in which the subunits forming the protein coat are arranged is always highly constant in any given type of virus, and is of a crystalline degree of regularity. In other words, the subunits are fitted together almost as precisely as are the atoms in a crystal of salt. The scale of the two kinds of crystal is, of course, very different; each of the virus subunits consists of at least one protein molecule, itself composed of thousands of atoms. Nevertheless, the regularity of spacing which is achieved in the two cases is comparable. In order for the subunits to maintain constant spacings and mutual relationships they must be held together in some way, and this appears to be effected mainly by weak linkages such as hydrogen bonds. It is postulated that each type of subunit must have a characteristic pattern of chemical groups on it surface, and that these are able to form linkages with suitably located groups on other subunits.

For reasons which need not be discussed here, there is only a small number of geometric arrangments in which it is possible to fit together subunits to form cylinders or spheres such as are commonly found in virus shells. The detailed molecular architecture of the protein component of viruses is restricted to a few standard patterns.

The next question to be asked, and which the electron microscope helps to answer, concerns the way in which the protein coat is put together when a new virus particle is made. As noted above, the protein molecules are regularly packed and joined by weak bonds. We now want to know how they come to take up their characteristic arrangement. Experiments on tobacco mosaic virus provide an answer to this question. If this virus is exposed to alkaline solutions (pH 10), the protein and nucleic acid separate from each other and the virus particles fall apart. The protein can then be separated in pure form. If a solution of this is made neutral, the protein aggregates spontaneously, forming cylindrical rods of the same size as the original virus particles. This can be demonstrated simply by examining samples of the preparation in the electron microscope. If the same experiment is done but with the nucleic acid present then this too is incorporated and complete, infective virus particles are formed. The nucleic acid, however, is not essential for the assembly of the protein. This discovery, that protein molecules will spontaneously assemble if the environmental conditions are right, is extremely important. It means that we do not have to postulate any additional agency to fit the molecules together. No template or mould, for example is necessary to give the virus particles their characteristic shape. Just as sodium and chloride ions, if brought together under appropriate conditions, will automatically build salt crystals, so will virus proteins spontaneously build virus shells.

It seems likely that most viruses are to a very large extent 'self-assembling' in this way, and that this is the normal manner in which they are produced inside cells. Essentially, virus multiplication consists of synthesis of their component molecules (protein and nucleic acid) inside the host cell, followed by spontaneous assembly of the molecules into complete particles.

These findings about the construction and assembly of viruses are significant not simply because they help us to understand the organization of an important class of disease-causing agents, but because the essential ideas can be extended beyond the viruses to many different kinds of structures found in cells. We may take as one simple example the flagella of some bacteria. These are extremely delicate fibres, about 150 Å in diameter and therefore visible only in the electron microscope. They occur either singly or in bundles, and they are responsible for the movements of bacteria, just as are the cilia or flagella of some plant and animal cells. Plate Id shows an electronmicrograph of bacterial flagella, prepared by the negativestaining technique. The flagella have a very regular beaded structure, each being made up of a number of rows of subunits which resemble those seen in virus particles. They correspond to single protein molecules or to small groups of them. It is known from chemical analyses of isolated flagella that they are made up almost wholly of one kind of protein. This can be isolated in pure form and, given the right conditions, it will spontaneously aggregate to re-form flagella. The parallel between these flagella and tobacco mosaic virus, both in the regularity of structure at the molecular level and in their self-assembling properties, is very close.

Similar studies have been made on other protein components of cells; for example, the fine filaments of striated muscle. The full significance of this work will become clearer in the next chapter, in which we shall consider what the electron microscope reveals about the structure of plant and animal cells. For the time being it is sufficient to recognize that with the aid of the electron microscope it has been possible to describe the structure of living organisms in some detail in terms of the molecules of which they are made up and the way those molecules are fitted together.

MOLECULES

It will be clear from the preceding section that some large molecules are directly visible in the electron microscope. They can be studied either by the negative-staining method, or by shadowing. From the biologist's point of view the most interesting of these large molecules are the proteins and nucleic acids. A considerable range of proteins has been examined, and these often turn out to be more or less globular, like those making up the protein coats of viruses. Many enzymes are like this. In these cases, of course, the polypeptide chain which forms the protein is tightly folded up. There are other proteins which are less tightly folded and which may appear long and thin in the electron microscope. The protein which makes up the collagen fibres of connective tissue is of this type. Nucleic acid molecules have also been studied, and especially deoxyribonucleic acid (DNA), which is found in the chromosomes and is the hereditary material of most organisms. This appears in the electron microscope as a long thin thread, rather like a piece of string, but only about 20 Å in diameter.

While the overall shape of these large molecules can be made out, the electron microscope reveals little about their detailed form. This is partly because the resolving power of the electron microscope is too low, partly because of the technical difficulties of preparing specimens in a suitable form. Perhaps in future some of these obstacles will be overcome; in the meantime, however, the study of molecular structure will continue to be carried out chiefly by indirect methods.

Index